Evolution and Memory

Caleb Gattegno

Educational Solutions Worldwide Inc.

First published in the United States of America in 1977. Reprinted in 2010.

Copyright © 1977-2010 Educational Solutions Worldwide Inc.
Author: Caleb Gattegno
All rights reserved
ISBN 978-0-87825-081-3

Educational Solutions Worldwide Inc.
2nd Floor 99 University Place, New York, N.Y. 10003-4555
www.EducationalSolutions.com

Table of Contents

Preface 1
Retention and Recognition 5
Somatic Memory 9
Perceptive Memory 13
Imagery 19
Cosmic Memory 23
Life Memory and Heredity 25
Animal Memory 33
Instinct and Memory 35
The Brain and Memory 39
A Mechanism of Evolution 45
Mutations 51
Games of Evolution 53
The Will 55
Animal Mental Dynamics 59
Discrimination and the Animal "I" 63
A New Kind of Being 65
The Uncommitted Brain 69
Education in Evolution 71
From I to Self 75
Awareness of the Will 77

Education	79
Human Evolution is the Story of Awareness	83
The Need to Know	87
Evolution in Man is Molded by Awareness	89
Freedom Explored	93
Evolution in the Fourth Realm	95
Mythologies	99
Horizontal and Vertical Evolutions	107
Every Civilization is Embarked Upon a Horizontal Evolution	113
The Creation of Language	117
Man's Evolution	121
Vertical Evolutions Seen as One Evolution - That of Awareness of Energy at Work	127

Acknowledgments

The original text of this book has been looked at carefully, and I must say even lovingly, by my friend and former colleague Caroline Chinlund.

She put a great deal of work in seeing to it that each sentence says best what it intended to convey. Readers are thus helped no end in coping with my proposals. My thanks go to her with my profound gratitude.

As usual my secretary Yolanda Maranga typed version after version of the manuscript with celerity and participation making the typescript ready for duplication. She too deserves all my thanks.

Preface

When writing the chapter on memory for my book "The Science of Education" I noted that it was becoming too big, too involved in matters that were very important but not quite obviously required by readers of the first treatise on the subject. Instead of cutting it out, I pursued the writing finding it easier to make a separate publication of these matters than to abandon the presentation of my findings. Hence this separate slim volume that scientists of education may, like me, find indispensable for their studies but that other readers of my large writing can be spared.

When the centenary of "The origin of the species" was celebrated in 1959, I was ready with an alternate vision of evolution. I lectured occasionally on it and talked about it with good willed listeners. It was clear that what I was saying did not seem an urgent preoccupation for almost everyone. In fact no one I met wanted to understand evolution in my way. So, I began to have monologues in silence and at odd places, thinking that perhaps I had gone too far in my not being carried away by the main stream. Reading here and there what was currently said about evolution showed me that the tools I had, though needed by others, were not at their disposal and I let myself delve more freely in these regions. The content of this book comes from such solitary and unrestricted excursions. These gave my ideas of the late 50's a chance to gather strength and evidence and they are here presented as I could not do at the time. I must confess that I am pleased with the result. But I am also amazed that my psyche in order to make me accept my vision of the current one, camouflaged their appearance in my consciousness as a contribution to a scientific study of memory and its place in our ways of educating the young.

Looking at all the references made to evolution in my writings on education over 30 years, I now find here and there a link of what is cardinal in this book: that individuality is the key to understanding evolution. But only here do I take full advantage of that awareness. So obvious and for so long hidden from any sight! This fact of not seeing the obvious has both helped me in becoming less easily swayed from finding what there is to find and alerted me to the possibility that the most trivial observation may be the germ of enormous progress. Writing this book has been a wonderful school for me. I saw some of my shyness go. I saw some of my resolutions such as not to entertain what I had not studied to my satisfaction enough, leave me. For example I have extremely rarely in the past referred to animals in my work and here I was trying to be fair to them in what they contributed to <u>my</u> evolution. I saw fit to speak of the most subtle thoughts I ever had and found in them tremendous promises for future studies and living. For that reason I kept them in the text in spite of the elusive nature of the content and the inadequacy of the language for their expression. Perhaps one reader will consider them.

As I moved to telling what decades of thinking after reading and after writing brought to me in this field, I found that the picture that was knit was both exciting and promising. Exciting because so many unexpected details became leading threads that open areas of investigation, promising because at once those leads yielded vistas that I will one day find had become major preoccupations for me and perhaps for others.

This is the main reason for publishing this slim text now. I believe it is needed, but I know it is going to help someone in getting more order in the inventory of the world's productions.

Retention and Recognition

For many students of man, memory is an essential ingredient of learning. Neurologists and some interested physicians, behavior scientists and computer programmers are among those who hold this view. This chapter will be devoted to a consideration of this idea and the proposal of an alternative.

It is quite a complex challenge to decide what we mean by memory because it covers so many aspects of mental functioning. It seems that memory tracks have objectivity and may have definite locations in our brain. People who have found themselves in certain exceptional somatic conditions as a result of using drugs, entering into certain states of trance or being under profound emotional stress have noticed "the reality" of their evocations was enhanced. Men engaged in electric explorations of the brain have noticed it is possible to trigger definite aggregates of connected images, leading them to the supposition that they have been figuratively sitting there in the neurons waiting for an impulse that would make them conscious.

The existence of an objective memory is made plausible by the useful and familiar computer. It is so much a part of our reality that the model of memory it suggests begins to impose itself as a method of explanation.

Still there is so much that we do not understand about memory. How is it that we forget either the tune or the words of a song when we learned them at the same time? How do we manage to "flash back" a symphony in much less time than it takes to actually play it? How does it happen that we can be simultaneously in the present and in a memory and know that the memory is sometimes "true" and other times has been altered surreptitiously.

We may have lumped together in the word memory a large number of different functionings that relate to our past and made it harder to study them because we apply approaches correct in the case of some aspect to other aspects requiring different treatment.

Let us call <u>retention</u> a capacity of a tissue to prolong the effect of an impact whether the prolongation is the result of an input from the central nervous system or a local one produced by the cells themselves. Retention is exemplified at the cellular level or perhaps even at the molecular level as we see it in chemical affinity, in the various tropisms of plants and other organisms.

Retention is therefore in particular an attribute of life and we must consider every human as a retentive system by virtue of the molecular and cellular soma that links humans and the rest of the living universe.

But retention does not work independently. It is integrated in a number of other functionings which alter at least its appearances. For example animals in pursuit of their prey have to learn to recognize alterations of their perceptions due to alterations in the backgrounds caused by light, the weather, or the presence of accidents in the terrain. Each impact from outside must be analyzed by those systems that relate the new contrast to what has been retained to yield some features which can be recognized as belonging to previous impacts.

Recognition is an important functioning and needs closer examination. It results from the ability of perception to start with a certain perceptible whole, ignore parts of it, and stress others while completing the whole only from the stressed parts. Recognition is therefore a power of the mind since it can use less than the whole to produce the whole. It is possible to stimulate the growth of recognition simply by finding the smallest part that needs to be perceived in order to generate each whole.

Somatic Memory

When we consider the somatic development that takes place in utero, the perception we envisage is proprioperception, that is, the network of feedback mechanisms that tell the self that the various aspects of the process of growth and elaboration of the embryo-fetus are working properly and towards the proper end. From such perception results the inner recognition that functions are working well or that there is a dysfunction which needs attending to.

After birth this proprioperceptive system goes on working. But a new element makes its appearance: energy from outside can intrude itself without consultation with the self. Coping with it is a new function which we all develop. In the prenatal states the self has integrated the past structures into the newly developed ones while subordinating the old to the new. The automatisms which emerge through this process indicate integrated wholes (structures and functions taken together) that stay whole through the direction of a very small amount of energy to energize the control mechanism. Automatisms are not a hundred percent automatic for they can be reached by some channels inhabited by the feedback devises and they can be made to alter their functioning. Some human disciplines allow people to regain the hierarchies of the controls and reverse their workings. It is important to note that humans are not limited to the study of their functionings; because they find the presence of the self with its capacity for feedback in them, they can also intervene in them.

At the level of somatic functionings automatisms are what we shall call <u>memory</u>. Memory therefore presupposes both retention and recognition. But also that what is retained is capable of alteration and integration, of working by itself but also when the self calls it in. Automatisms are the first forms of memory which will develop in time as the self on one hand objectifies itself in different forms and on the other hand manages to go beyond the objectifications. The somatic structures both form the support of the objectified self and are permeated by the energy of the self giving them complex functionings that are displayed in and around them. Thus the self can reach the locked up energy of the structures as well as the residual energy present in them and in the interstices. What the self does in time is translated into either structures or dynamics. The soma is both: a reflection of the self in prenatal conditions giving itself a somatic memory in every one of the somatic integrated wholes that can summon themselves when the self needs them. The brain's structure participates in this too. All the integrated di-encephalic functions are part of the somatic memory which functions exactly as we shall find the postnatal memory to function.

Forgetting, in the prenatal stages, means to have become automatic so as not to need recall into consciousness. Unless they are forced by the self into consciousness, automatisms do all their jobs leaving the self free to work on new challenges and making the past into an instrument for the self. So most of us have forgotten (in that sense) all we did for months and months in the prenatal and postnatal phases but we hold a total memory of those times in the functions and the structures we produced during those months.

Perceptive Memory

Upon that somatic memory we build our perceptive memory soon after our vital vegetative memory has been automated in the first few weeks of our life ex-utero. By swallowing, ingesting, digesting, assimilating, evacuating we have given ourselves the know-hows that are part of our present memory, the remotest past of our postnatal existence. We all "remember" all these functionings all the time by knowing how to carry them out every time. Our perceptive memory presents two aspects. One, similar to the somatic one, is the result of knowing in terms of energy what reaches the soma, and one, very different from that, is concerned with a secondary form of energy which does not identify itself as quantity of energy but as the distinctive form it takes for which the soma has developed sensitivities.

If the soma only received quanta of energy, perceptive memory would be somatic memory. But some concomitant factors depending on arbitrary conditions (such as for example, the variable areas of the retina that are lit and say, are a function of the hour of the day) create a situation where the impacts are no longer pure energy impacts. The self aware of what it receives distinguishes both (the amount and the area) and holds them as different and as calling-in the self in two different ways.

To know the energy of what we call the red photons, is to assess the amount of each quantum of energy received, determined by the frequency . The brain will remember red photons if it is capable of recognizing its own transformation when such quanta are added to it. We know physics instruments that can do it. Our substance can too and there is nothing to wonder at if we all remember colors. That working of perceptive memory makes it only a continuation of somatic memory. But since the names of colors are arbitrary (they change from language to language), to remember that arbitrary component is a different matter demanding different behavior from us,' as we shall see further in this chapter.

Because the self can perceive the objectivity of energy and the dynamics of energy in itself, we shall be able to understand much that has appeared mysterious and has created much confusion among students of mankind when they did not resort to the presence of the self in knowing. Our sense organs are exposed to energy impacts from outside. These organs are specialized ends of sensory nerves that take the inputs to regions of the brain in which preexisting connections can be activated. When the organs receive impacts there are objectively distinguishable attributes of these impacts, such as in the case of sight, intrinsic energy levels of the photons themselves, the number of photons received, their distribution upon the retina, their duration and the variations of their input upon that duration. All these the brain can distinguish and does, if we know that the self, capable of awareness, dwells in it. The impacts being energy increases are retained as energy (i. e. not leaked out). The way the self acknowledges that the impacts have been retained is by becoming changed by them. "Retained and recognizable" is equivalent to "remembered." Objects reflecting photons that reach our retina generate in the areas lit on it the equivalent of (what will be called one day in English) their "shape." Shapes are therefore objective attributes of objects. We can remember shapes as we do colors since they can be retained and recognized because of the energy inputs they represent. The property of shapes — we shall one day call in English "similarity" — is also objective in that the brain involves nerve fibers whose arrangements are one to one correspondences of the impacts on the retina, themselves (through the optics of the lens) one to one correspondences with the "points" of the objects. This property, a point to point transformation, gives perception the possibility of knowing a class of impressions as defining the class that includes all the impressions. Shapes and similarity are part of the same perception because we are at once conscious and mobile, and our mobility imposes upon our retina variable amounts of energy which can be organized through perception. This makes it possible to recognize as one shape a class of impressions of shapes and to remember one of these impressions so as to trigger the class if needed. The memory of which we are conscious, is made up of one impression and the transformations which would produce the class.

Because of this babies are at peace with all moving objects and, unlike cameras (which record only the optics of the situation), do not think that their beloved ones actually get smaller when they move away from them, or vanish on the horizon, nor think that when they bring their hand close to their eyes it gets bigger.

In the universe around, objects and transformations contribute to the making of reality, so long as the energy impacts can be actually processed by the self once they are received by the self and its instruments.

Remembering is as much a process of retention as of recognition and we hold classes of impressions and their transformations in our brain much more easily than individual impacts (which is what was conceived by John Locke and early epistemologists). We recognize varying distances as much as we do objects; changes in hues as we do colors and our language is proof of these sensitivities. Because of the dynamics ever present in our self and its objectifications we can find ourselves at home in a universe which is in perpetual change. A rigid photographic memory would be the worst way for our organism to adapt to the dynamic environment we find ourselves in. That is why we do not have it and when one of us is found to have an approximation of it, we single him out for examination as a curiosity. Observers of the functions of memory have noted how much intelligence eases retention, precisely because the retention in question was of single facts that were to be kept in isolation, (as for example the names of rivers one has to learn for geography lessons). Intelligent people used mnemonic devices and other tricks to ease retention.

While we are not worried that we shall forget what we can re-invent, we do worry about how to remember that which is not part of a system. Remembering has here the meaning of non-functional retention, requiring a great deal more energy to exist than retention of what generates itself because of inner dynamics. Energy must be held at the service of retention in two ways. If we use outside energy — as when we get the photons reflected from objects — we do not have to consume the energy of the self for more than the production of inner dynamics and we therefore remember easily what we associate with that energy and the resulting knowledge resembles the somatic and perceptive ones. But we can also mobilize the free energy of the self and give it a degree of immobility that makes it distinguishable from the rest of our energy at least to the self engaged in the perception of inner life. To understand this process which goes to form both memory and the awareness of it as knowledge we must examine the process of imagery which is an activity of the mind. Mind in this case is taken to reside in the soma extended by perceptive memory.

Imagery

It is easy to realize that energy from the outside reaches the organs created for its processing, i. e. analysis at the organ level and transmission through the nerve fibers to the centers where so many neurons are available for its reception and retention in the form of chemical changes. These chemical changes are recognizable to the somatic sensitivity placed there from the start for such functionings. Hence the self is informed that they took place and it can reverse them because it is energy, knowing itself to that extent in that location. Sending such a-mounts of energy (from the self through the nervous paths) will activate those parts of the sense organs that reproduce the impact of the original energy received from outside. This we call an <u>image</u> and the process, <u>imaging</u>.

But because all this remains within the somatic system the self alone can know it. Because images are part of dreams and imaging is more common in a state of semi-consciousness or daydreaming, we can safely place the process in the diencephalon and understand why images as energy, and emotions as energy, are so often experienced together. The self is aware of both and of either at the same time so that it can tell itself whether it is considering their co-presence and their distinctions. In nightmares the intertwined pair causes an upheaval but the self knows whether the images or the changed humors are being concentrated on. In sleep-walking the images and the reality are exactly congruent and one can move within the image as one does in reality, knowing one to be the other and mobilizing the other functionings involved as perception and action do within the integrated functions that involve the past as memory.

At the level of imaging, the self uses the sense organs in reverse. To manufacture impacts which in turn send a message to the brain to receive the confirmation that the image and the original impact have enough in common to be assimilated. But clearly the amounts of energy are vastly different. An experiment will convince any reader at once.

Select any object sufficiently well lit to be seen distinctly. Close your eyes, it is still possible to evoke the object. In order to be sure that it is not a phenomenon of remanence keep your eyes shut for a while, imagining whatever you want and then, still with the eyes shut, evoke the object once more. Opening your eyes to see the object you can at once note the difference of intensity between the retina activated from within and the impact from added energy. The only immediately noticeable difference is in the intensity or amount of energy, the other perceptible components are present in both the images from within and from without.

Images are energy molded by the senses. Since that energy is part of the self it can be sent from one organ to another and be affected by this dwelling in each one. Images can be described as having dimensions resulting from their successive restructuring due to the impacts of the senses. The same energy will bear the imprints of its various journeys and can be known to the self through these imprints. A baby can hear a voice and evoke the soothing visual image of an approaching loved person; hearing steps on the floor we can all evoke a person as if he or she were being seen. The content of the mind is filled with links between the images just as our brain is filled with connecting neurons. Although they are less visible, they are not less perceptible from within. Our everyday experience will supply us with as much evidence as we could wish for of those components that are produced in the soma by the fields which affect the content. Think for example of a familiar name, or a shop and note the images. We can become aware of associations as easily as of images, of transformations as of forms. Our memory not only contains associations retained from past experience but can make new associations by its own dynamics. The self can also reactivate material that has been retained and find in it elements that have escaped it. One can transform one's past because the self dwells in one's memory. From this it follows that we have another task, that of understanding how retention can be made to be faithful to the original impacts. Or is it ever?

Cosmic Memory

In the atomic and molecular universe, energy is quantified and it becomes matter in such a way that we can say <u>it</u> knows which systems can persist. Change is defined against this frame of reference. In that universe there is a coherent physical definition of sameness. Chemical affinity prescribes which wholes can and cannot be formed, which among them can exchange their components, which can last and which are unstable, which are excluded. As if atoms and molecules with their nuclei and electrons knew exactly what to do and did it without doubt. Changing conditions affect in precise ways what may happen. All this is the basis of retention and recognition in the molecular universe and can be called <u>cosmic memory.</u>

Because we are molecular, cosmic memory is ours and we display it. Because we are also members of the realm of the living, which on earth has taken the form of the cellular, we see cosmic memory subordinated to <u>vital memory.</u> Reproduction makes one assemblage of molecules capable of reaching a certain complexity and then produces two samples from one followed by a period of directed growth to the point where the sub-division takes place and so on. Such a process had been tried out already in the molecular realm. At the atomic level the nuclear edifices reached a certain complexity and then spontaneously broke down indicating the boundary of what matter and energy can do together through the same processes. Radioactive substances and the heaviest stable atoms are the demonstrations of what compatibility is in the realm of the cosmos. Although matter and energy had to wait for man to tell their story, they lived it to its end — and perhaps the dramas of matter going on in the cosmos not yet revealed all their potential. Molecular existence has produced its own "bottle necks" and "dead ends." To find a way out we must move to the center of the Table of Mendeleef which provides a new cosmic adventure on earth, an outlet in the large molecules based on the atoms of carbon (and also of silicon) that can form almost endless chains. But there again, growth is, so to speak, doomed. There too complexity can go only so far and then the edifice breaks down. Both attempts in the molecular realm did not meet the challenge of indefinite growth, although in the first the direction was nuclear complexation and the second electronic concatenation. In space-time, energy alone (and its objectification: matter) could only produce stars and their clusters in which man today reads an "evolution" and "a life" that has taken billions of years to become what we see today in the cosmos.

Life Memory and Heredity

Life is an attempt of energy to do what was not possible in the molecular realm: to produce an endless evolution. To do this there had to be several points of transfer from the molecular realm: 1) large molecules only possible with some atoms were so to speak adopted, 2) the use of catalysis as one of the ways to economize energy and matter so that particular molecules serve to facilitate particular syntheses*, 3) spatial organization of matter so as to produce with the same components, differences in the properties of the results, 4) maintenance of the molecules in contact with each other to produce a conglomerate that remained enclosed in one cell, thus making it possible to work out syntheses that could perpetuate themselves with the minimum amount of energy and of matter.

Because of the success of such conglomerates they are still, like the atoms and molecules, at work in that part of the cosmos called earth (and most likely in other places as well).

---

* Such catalysts known as enzymes only recognize some structures and produce some transfers so that we can say that they have a memory of these structures and processes but are insensitive to all others.

While growth and breakdown followed by growth followed by breakdown and so on, exists already at the molecular level (and could serve as a sketch of what memory is by determining both at which stage to breakdown and which fragments to produce) in the cell the process of breakdown takes a new form. Each molecule of a certain subset of those contained in the cell, becomes two by producing its double. Special catalysts and special conditions within the cell make this process one of those possible — where selected matter from outside can get in and be used for the syntheses. Instead of a duplication of every stable system as in the case of the large molecules that do not breakdown, in the cell the process is timed and forms a hierarchy that goes on repeating itself unless there is no matter or energy available. In such cases it stops and we say that the cell dies.

We have a choice at this point between understanding <u>time</u> to be a form of the universe, as Newton, Kant, Einstein, among others did, or observe the fact that duplication of molecules produces a structuration of the same reality which is recognizable by change rather than by persistence, and say that time is generated by a sensitivity to that change and those structurations. Each system then has its time just as <u>its</u> matter will give it <u>its</u> space. Energy could then be a space-time synthesis, as would process. This would force us to acknowledge the primitiveness of all the fundamental instruments such as: energy, matter, process, space, time, sameness and difference, and accept that these are necessary awarenesses already at the level of the cosmic and the vital happenings. At that same level we found retention and recognition, and therefore there we shall see memory coming into being, the memory that organizes time so that individuals that display properties already existing can be produced in space-time. This does not preclude other processes which produce what does not yet exist, – the new!

That so much of what is successful in the history of the universe can be stored at the molecular level, (that DNA by exploiting the capacity of structure to be complicated and of process to be triggered in precise order is capable of commanding the perpetuation of species and the production of mutations) – only tells us how thorough the working of evolution has been from the start. Molecular biologists have been inebriated by their vision of so much happening at such a level, but the rest of mankind has remained skeptics in front of their enthusiastic claim that all is in the DNA.

For our purpose we find memory at work in the universe in harmony with variation and with the uncommitted, at every moment allowing all phenomena to be knit together without claiming to be more than it is: contributing much but not holding the commanding job. Memory is proof that some of the processes in the universe have found themselves adequate and integrable by the new, facilitating its tasks.

Memory, like many functionings of the self, must be seen as part of evolution. If it frees the self for future adventures, it is biologically acceptable. The holding of the past is an instrument of evolution if it does not hamper the future. Thus it is possible to look at the vegetable realm and see that 1) some forms are viable on earth because they go on surviving for millions of years and 2) some other forms are attempting the new to test viability.

This is the meaning of evolution within a realm that does not seem to require the disappearance of what knows how to survive but can also be transcended.

The collective past is maintained by the processes which have been tried out in the universe (or on earth in particular) and found to be compatible with the existing varying conditions. Such processes as the withdrawal of some activities seasonally (as in deciduous trees contrasted with evergreen trees); separation of sexes or anatomical or physiological separation of reproduction; flowers digesting insects and contributing to what the sap does in almost all other plants.*

On our planet variation goes on at the same time as perpetuation. The survival of the fittest is more an instrument for thinking than a description of the content of nature. As soon as subdivision of cells stops leading to two individuals, and two cells remain attached – say by accident – instead of separating, the new has a chance to test itself. If it is viable, we see multicellular organisms appear and along with them their universe produced by their being. All that which could be continued in the new situation goes on and can be carried on to a next generation if it becomes a function that can be summed up in the transformation of the DNA, as previous changes have. The molecular continues to be present in the vital and to pull its weight.

---

* This last attempt is one of those where we can see the breakdown of a sequence of adventurous moves. The unsuccessful development stops at what remains viable but is rather a curiosity on earth.

New organisms always represent a proof that a form can survive if it takes care of integrating what has been tested and found viable while subordinating it at the service of one new functioning for which as many formulations can be proposed as are compatible with it. Thus once leaves appear they are susceptible of becoming the object of variation, and many kinds of leaves result, each attempting some special way of being a leaf under the prevailing circumstances. Some have hairs; others are covered with wax; it seems all shapes and sizes are tried out and a great many are found to be adequate for the functioning of leaves and so they survive, as well as they enable the organism to survive. The realm of plants includes bacteria as well as sequoia trees and douglas firs. They can be conceived as belonging together only if we say they are cellular organizations of the molecular with one additional common function which is that they manufacture their substance synthetically from the molecules on earth. Plants display a knowledge of handling energy which can be seen already in the less complex monocellular organisms. This specific way of processing molecules not found in the cosmos (here defined as the realm of the molecules that remain molecules in their interactions) we shall call <u>the vital.</u> It permits the forcing in space-time of 1) certain reactions to take place one after the other (instead of letting a product be its end) and because a certain molecule is present in the confined space and meets others and 2) recognition that the specificity of reactions with specific reactors and specific catalysts as are present can be translated into a programmer that perpetuates the organism in certain circumstances.

Scientists have known the realm of the vital through the millions of species studied over the centuries. The realm of the vital can create its own kind of memory and from generation to generation take the attempt that each organism represents, to a new form that is tested for viability. The work of the vital is incomparably more complex than that of the cosmic. Part of it is still beyond our comprehension and may challenge man for a long time. How can the sap reach the top of redwoods so much further from the ground than a natural pump can supply? How can so much chemistry of such a complex nature take place in a bacterium with an almost infinitesimal expenditure of energy? These are lively questions for engineers.

In this book what we shall retain particularly from the realm of the vital is a new way of handling energy that we see at work. Memory will be concerned with it. Each organism will receive from its parent or parents the gift of working at a certain level and with a certain efficiency. The job of every individual life will be either to adapt itself to the circumstances or to do that while attempting to expand the given. Perpetuation of the species and mutation are inscribed in the fact of being an individual. Memory serves the first but does not prevent the other. It is in fact the economical way of maintaining a given within the circumstances of change. It makes available to the individual the proven work of generations but not as a static transmission aiming at identical repetition of functions from parents to offspring. This would be untrue of a variable environment where no one knows what will happen. Individuals are freer to adapt than groups or masses. Adaptation means readiness to meet change. The way in which an individual meets change must be compatible with what is transmitted in the molecular and vital heredity. The world is full of such successful adaptations, i.e. transformation has been inscribed in the living from the start. Permanence is an appearance defined by the ignoring of change. Age or the passage of time forces the cancellation of such ignoring and change is stressed, then remembrance, reminiscing at, as the corrective to bring back the lost stressed permanence. In some trees, the rhythm of seasons is the substitute of reminiscing, each spring a renewal of growth begins but only from where things were left the previous fall, the memory of the past being what one is now, permitting another lease on life. A tree can grow forever unless its functionings are interfered with when the wind uproots it, or lightning strikes it, or drought or fire change the environment.

In fields, by contrast, the seeds that grow into the grains, the fruit or the vegetables, produce from the cosmic conditions (molecules and radiation), that which will permit a new realm to develop on them.

Animal Memory

Once again, it is an almost superficial transformation that begins a new adventure on earth. Rather than synthesize from the cosmic elements with the know-how of vital energy, the new organisms break down what they find that is living in the environment before they resynthesize what they want in order to make their substance or renew it.

<u>Animals</u> are the labels for these organisms. They can be one or many celled. Their functionings are freer than those of plants. Because they do not have to use part of their energy to bind elements together, they begin beyond the point chosen for and maintained in, the realm of plants.

Animals do not need to have memory of the syntheses which plants have to remember to produce individuals of certain species using their DNA and the environment. Animal cells are space-time locations where chemical reactions take place but after two jobs have been done. After an animal ingests an organism, the molecules of the food organism are separated into 1) those that are not retained and which are excreted (by a variety of processes) into the environment and 2) those that are retained which are first broken down and then all or some are processed, (the rest is again excreted) to become molecules compatible with the present content of the cells and needed to provide energy or matter. After that they become part of the matter that moves from some cells to all cells (if we consider polycellular animals) and enter into specific chemical reactions. In the animal realm, synthesis follows analysis. Each species in particular begins by ingesting particular organisms that provide the level from which these animals will take their leap. Animal energy is going to generate a new universe simply because it does not do the job of cosmic energy which produces matter, nor the job of vital energy which produces plants. It begins where these leave off, rather where vital energy left off in that it integrates some plants or some animals. What animal energy does is to dwell in an organism that has gained the freedom of movement, seeking its sustenance where it is. Each animal knows what is its food, that is, it recognizes the tissues that it can process and store. If it does not get what its functionings require, it seeks for more. The minimum energy that is required is that which compensates what the organism consumes to maintain itself. Below this level the cosmic law of thermodynamic applies and the organism degenerates. This is called death when the chemical processes characteristic of that organism have stopped and no ingestion of other organisms goes on. In the finite universe where animals dwell there may not be enough organisms to go around to enable those which integrate them to go on. The basis of death in the animal kingdom is the disappearance of some organisms so that others may survive until they themselves serve as substance for others.

Instinct and Memory

With the extra energy saved from some of the syntheses and that gained from the analyses animals generate a new universe: that of behavior. Behaviors are snapshots of what form plus a certain amount of energy can achieve. In the here-and-now there are behaviors. An animal is the creator of a constellation of behaviors. All those animals which create similar constellations form a species. The whole animal kingdom is needed to explore the possibilities that forms are free to express.

Since it is the surplus energy available – surplus because an animal does not synthesize its food, that makes an animal capable of behaviors, we shall call <u>instinct,</u> animal instinct, the energy that goes to set in motion these certain constellations of behaviors. Instincts can be looked at as memories between generations, making possible the perpetuation of the past in its compatibility with individuality and adaptation.

When we come to animals integrating we see evolution at work on a more flexible instrument, since animals integrate the work of vital and cosmic energies at the level of instincts, evolution needs only to extend the constellation of behaviors to generate a new animal. A large number of varieties is possible within one species and one instinct, the more complex the species or the instinct, the easier to produce a variation to distinguish individuals by their own features.

Animals have only their instinct to store in their memory. Beyond this the individuals have experience which does not affect the hereditary mechanism. There is retention of experience only if it is compatible with the instinct, and the memory of individual experience can no more be transferred from one animal to its descendants than from one animal to others of its own generation. Animals learn what their instinct permits and remember during one life what they learn. Mutations occur in animals as in plants when energy affects the cosmic basis of energy transmission, and the DNA branches are changed by accident or selectively. The mutation that is produced keeps grosso modo the new animal within the species but opens the way for a new species if a sufficient number of alterations take place – perhaps over a number of generations.

In the animal kingdom the force behind evolution is in the capacity of the individual experience to affect the instinct which then affects the vital and cosmic basis of heredity. The cumulative effect of this alteration of the instinct leads to instincts that are distinctly different from the original ones. With this change of instinct may go some change of form which has to be made to be adequate to the expression of the instinct. A certain animal can be seen to be the form that expresses best an instinct, that makes possible the objectivation of a set of behaviors. When each animal is seen that way, it is easy to see why the most adequate expression of a soma is an assemblage of tissues that are integrated in a center of command that is the nervous system of that animal, a center directly activated and maintained by an instinct, that is, the energy supplier of the soma.

The animal kingdom can be looked at again, this time in terms of the subordination of all functionings to an instinct that integrates the corresponding cosmic and vital objectivations via one organ and its extensions. If the instinct dwells in the nervous system from the start it is clear that whatever the unfolding of the somatic development, the result will be an animal of that species. During its individual life the instinct will be expressed via its objectivations and its functionings but as it can only have an individual response to varying stimuli it will produce one actual animal out of a spectrum of possibilities. In so far as conditions and responses do not vary substantially, the instinct will express itself more or less in one way through all the members of the species. But in so far as they can vary (and considerably) the behaviors will affect the instinct and subsequently its expressions. A new species may result. No need for the hypothesis of the survival of the fittest to understand the present animal content of our planet.

The Brain and Memory

The appearance of a nervous system in animals results from the overall process of integration and subordination which comes about because of the realms involved in animals and their proven ways of meeting with energy the challenges of energy to energy. The nervous system is an economical approach to having at the same time 1) an instinct that gives itself a form to express itself in the cosmos (here, our planet), form which integrates the vital and the molecular and 2) an individual that uses the form to test its adaptability to the prevailing conditions.

The nervous system is brought into being soon after the soma is begun. It keeps pace with the soma and is always adequate for it. The nervous system of the completed soma is made up of level upon level, each subordinating the previous one by integrating it. This whole process is at the same time in harmony with what the DNA brings and the individual does with the energy and matter available in the environment. We therefore end up with an animal whose instinct can immediately express itself in the outside environment as it was able to do in the mode of maturation of the offspring that the instincts of its species had adopted (oviparous, viviparous, spores etc.).

Still there is nothing in that form of maturation which prohibits growth from going on once the animal is in the environment. In fact, the individual will use its functionings established in that period to adapt to the environment in order to continue the use of those functionings and of some new ones that stem from the demands of the environment upon the individual. To maintain the command of the instinct upon the functioning soma more nervous tissue may be called in and is called in; the newest layers which come into action according to the need for new functionings that may be new for the species in the present circumstances. Penfield speaks of "uncommitted brain" and suggests that its quantity increases as the complexity of the brain increases, particularly among monkeys, apes and human beings.

The new kind of evolution is characterized by the way animal energy or instinct, becomes the domain. It is no longer the mobility of the form – which in the origin made possible the finding of ready-made organisms to be assimilated – that characterizes the animal kingdom. Instead it is the capacity to affect the instincts which in turn affect the form while somehow enough residual energy is left to allow individuals in the species to use their individual experience to affect the instincts so as to integrate the new by subordinating the old and so on. Thus when we observe animals (i.e. individual animals) we find at work a multiplicity of directed behaviors partly under the orders of the species instinct and partly by the processing of the impacts from reality. No student of animal behavior has failed to report the work of animal intelligence which shows itself every time routine is insufficient to cope with the present challenge. Naturalists seem astonished by this finding as if they started with the preconception that instincts were rigid and animals could only display collective behaviors representing these rigid instincts. Individuality permits deviations and adaptations through innovations. Innovations are the expressions of adaptation to begin with, forced upon the individual by circumstances. Later on in one individual's experience, a number of innovations may lead to the recognition that their instinct is compatible with a spectrum of adaptations and suggest to animals some of these clever ways of operating (say while they are hunting) which astonish naturalists. Intelligence consists in the capacity to call in some of the available functionings which have not been spontaneously triggered by the encounter of a challenge. Intelligence is an animal functioning because it works in harmony with what instincts have been doing all along in the life of each individual.

We therefore see an individual's memory working through the instincts and bringing to the individual what the species has achieved, first through its evolution over millions of years and second, through what each individual has learned in its own life. We find the first imbedded in the soma, the functionings and a vast number of behaviors and the second through the few (or many) behaviors that single out the life of that individual because of its individuality. This singularity expresses itself in the use of the newest layers of the brain while the other layers concern the species' fundamental functionings.

We have met evolution in three realms and seen it to be concerned with the functioning of energy. In its cosmic form it has produced the universe of atoms and molecules which own the properties of retention of extra energy and of recognition of what makes other atoms and molecules capable of being assimilated or shrunk from. The simultaneous functionings of these two properties makes what we called memory and we found it at work in the cosmic universe.

In its <u>vital</u> form we have seen evolution use cosmic properties to produce an edifice of cells, organisms that could envisage with economy the immense variety of structures that form the vegetable kingdom where synthesis is the process of creation. Retention of form – structure plus compatible dynamics – and recognition of what can be done through its connection with the environment allow the vital to unfold and display in the universe a hierarchy of viable alternatives which <u>when successful last forever</u>. Memory in the vegetable kingdom is coextensive with the permanent presence of what each form can do to maintain itself on a residual energy affecting molecules. In the vegetable kingdom, memory is almost as extensive as all functionings taken together that make variation by individuals less of a force of evolution than adaptation to the conditions. The changes which generate new species come from those factors which affect directly the processes which produce the actual form out of the <u>material</u> in the environment. Plants being so close to the cosmic substance of our planet reflect the cosmic substance of their localities and show their vital constitution by doing with this matter what is compatible with the inherited functions.

In the animal kingdom this last feature is still visible but since it is organisms and not the cosmic substance that are assimilated, the determination of a species is less directly dictated than in plants. In the breakdown of the ingested organisms there are varying stages (from the whole to some molecules or to their constituting atoms) and individuals in certain environments have a latitude to proceed in different ways, more easily than plants do because the substance of plants has to be reproduced from scratch from the molecules found in the given environment.

A Mechanism of Evolution

Because animals' evolution is the story of how instincts can be affected by individual behavior, the display of how one energy – that is in excess of what is required for reproduction and for maintenance – can generate clusters of behaviors is the history of the animal kingdom. To look at animals is to look at what they do individually – and in certain groups – with their endowment to change survival and reproduction into the unique constellations of behaviors that fill their universe. It seems that evolution in that realm can be viewed as the experimentation by species and individuals to find out what can be done with certain endowments when these are coupled with a capacity to adapt deliberately to what is compatible with them.

Tropisms are plant behaviors and are predominantly directed from without. In animals behaviors are directed from within and give the impression of being in part willed. A spider must have in its heredity how to make a web, but in order to actually make one the demands of the place it has to hang on must be integrated with the instinctive behavior. Spiders learn to use the hereditary to produce the actual and do not find in their DNA what the DNA cannot contain – an <u>a priori</u> survey of all crevasses in the rocks where they might stretch their web, for example. Their behavior is a blend of that of the species which contains hunting with webs, intelligence of the space selected and of the mechanics of hanging webs and the capacity to learn from mistakes within that frame of reference; to repair broken webs. Each species has to learn beyond what the instinct transmits and that is how the individuals can affect the species. There is no need for the idea of transmission of acquired characteristics in order to understand mutation if what is needed is the notion of individual flexible adaptation. The job of the DNA is at the level of the form and of the functionings which permit individuality to take over to allow adaptation to that which cannot be forecast and programmed. By distinguishing instinct from experience we have allowed the recognition of a permanence through instinct and of variation through individual experience.

The animal kingdom is the domain of all possible behaviors, and the study of how forms permit behavior, affect form, is the field of zoology. A hierarchy in the animal realm cannot be a linear sequence of structures evolving one from the other. It results rather from the temporal imprint of what individuals can do in certain circumstances to what was given them and shared by others. There is a place both for DNA and for experience in the succession of generations. But since the species, once produced and proven viable, will go on to display their behavior for millions of years, we must consider the transmission of the bases for some behaviors from generation to generation. This can only take place at the cosmic-vital level. Compatible with this basis (which is also part of the molecular and cellular realms), will be the individual alterations of those molecules that are accessible to the energy impacts in the individual. In the way the form in the vital affects the cosmic, the behavioral will affect the vital which then affects the cosmic. Experience is not an attribute of a certain age. At all ages it exists and utilizes energy. When only a small number of cells is constituted, that energy is in close contact with the content of the cells and can easily directly affect the molecules in the DNA. When there are many more cells already produced, the same effect is possible but via the channel of specialized nervous tissue. Up to the stage of independent behavior, the individual is translating experience of change into awareness of what the form can do; once independent, it is awareness of what the form can be stretched to do. In this learning the individual is not simply compelled by its instinct but recognizes what the instinct makes possible and suggests. Practice is needed to take oneself beyond the experience of one's limitations. By concentrating on some behaviors and bringing all the free energy to bear on them, one or more individuals in the species can push beyond their ordinary boundaries. According to the extent of the deviation, an animal remains in a species or generates a new one with a form corresponding to the new behavior. This form being cosmic-vital can affect the DNA and in the new organism the extended behavior will be hereditary.

Beaks of birds, wings, claws and feet show how within a group tiny variations in form due to individual adaptations have produced hereditary features. The same sort of variations are present in pure behaviors such as nesting or migration.

All individuals have uncommitted energy available directly or through their nervous system. Even in socialized groups like ants and bees, there are deviations from collective behavior simply because the state of individuality exists and makes that possible. The uncommitted energy allows the possibility that one individual attempts a move compatible with the instinct and the task in which it is engaged. The result in the circumstances may make enough of an impact upon the individual to force awareness that what was attempted can be done again and again. Such behavior because it is compatible with the instinct of the species can be adopted by other individuals and become second nature. At that stage the DNA is affected and the behavior descends into the offspring whose new instinct will mold the form during the period of gestation. When these offspring are still very young, their behavior may differ substantially from that of the offspring of the mainstream of the species that was unaffected by the new behavior.

The two important points in this are: 1) that evolution takes place at the instinctual level first, it is energy affecting energy and 2) that energy compatible with the form taken by this instinct in the species affects "the point of entry" of the instinct into the process elected by that species to produce its somatic form. If it is through a modification of a molecular branch in a large molecule, all the mechanism already exists for doing that and has been tested over billions of years by billions of billions of individuals. But it may not need to be through such a process. It may be sufficient that a potential (an electric potential) changes at the molecular level, and only locally, at an early stage in the unfolding of the process of construction of the soma, for a change to be sufficiently powerful to generate visible consequences. It maybe that the effort of reducing everything that can happen in the triggering of the passage of life into an individual that has to make itself, into one model proposed by one scientist in the field of heredity is the cause of our confusion when facing the alterations of species on earth. May it not be more useful to look at the relationship of individual and species differently than as if it were a "printing process?" where species precede the individual and all the individuals are identical?

Mutations

If energy can become matter and matter becomes forms, energy can affect form as indeed is shown everyday in physics and chemistry laboratories. If we are concerned with <u>one</u> molecule in the egg at the start of living specimens on earth, a large molecule with many submolecules aggregated in it, the presence of the instinct in that space manifests itself as determined potentials that can be altered by alteration of the instinct. These potentials can alter the mode in which the molecules behave. The perpetuation of a species results from the ability of potentials to reproduce themselves at the moments of subdivision of the cells and that will permit the development of the individual within the species. If certain of these potentials are selectively affected by changes in the instinct in the parents, the changed instinct will dictate a change in form: a mutation. This will be functional if the change is viable, and if it is not viable, incapable of adaptation as in the case of some "errors."

In the vision of evolution sketched above time becomes a major figure in the drama. The individual's actions involving energy transactions affect the expression of an instinct enough so that the energy of the instinct affects conditions in the initial cell and the initial hereditary material that will produce one individual. We are concerned here with discrete individuals not species. Individual changes that are viable, if they are repeated often enough, produce a group showing distinct behavioral characteristics and are called a species. A new one.

In the process of reproduction the potential of the individual's instincts can work better in the space-time of gestation if the offspring has a longer time in which to receive the impact of the changed instinct. Animals that take longer to have their offspring produce fewer individuals. But if a couple has managed to affect the instinct for one particular behavior, they will endow their offspring with greater ease for that behavior since the offspring will be supplied with the vital material from the mother. This material in the blood of the mother is produced by some interaction of instinct and matter from the environment. The mother can develop a taste for an organism in the environment and force the embryo in her to either change the material received to make it what is needed for the development to go on as it goes on in other members of the species, or change itself and use the material to make new molecules which will affect the form of the soma.

Individual variation cooperates with the hereditary procedure but produces new individuals capable of further variation.

***

Games of Evolution

In animals whose gestation period is longer, there is also a longer feeding period which gives the offspring the freedom of readying themselves for independent activity in the environment. In these animals we see how the constellation of behaviors becomes more numerous and complex. The result is that variations are also more numerous and cover a wider spectrum. The notion of species becomes looser and even not very helpful. What is stressed now is individual living, even within the group.

While parasitism and symbiosis have been tried out at the plant level and found viable, animals have developed another form of cooperation to increase the efficiency of their living on the environment and provide themselves with time to spare from feeding and reproducing themselves.

Already young offspring invent games which means that not all is learned before one is born as heredity could have done and may still do in far less complex animals. A certain period of life after birth is dedicated to learning in the environment. Because parents cater to their little ones, the offspring have time to spare from the struggle for survival and spend it in acquiring skills which are compatible with the instincts. Under the vigilant eye of the grownups the little ones quickly discover themselves and learn by experience. One of the roles of parents is to place their offspring in situations which are challenging and from which they can learn. Sometimes the parents associate themselves with their young ones as targets and playmates, occasionally showing their strength or their impatience to give the offspring a sense of their social boundaries.

This period of play and learning in a family or clan is found in many variations, throughout the animal kingdom. This makes it clear that it is not the result of chance or accident, but of the existence of individuals, who are capable of affecting instincts, and of evolution finding one more way to use variation to produce the radically new.

With the advent of a learning period, behaviors become the field upon which energy will be acting. Instincts and the preservation of the species are no longer central. The generation of new constellations of behavior becomes the purpose of the animals which see the environment more as a set of opportunities than as the ultimate molder. More and more of the environment is explored and studied by the animals with largest brains and of their life does not need to be entrusted to heredity. There is room for education of the young even if adult education does not yet make an appearance.

The Will

Evolution has taken a liberating turn. It now manifests itself in a companion of the instinct, a <u>will</u> capable of holding down instinctual reactions and of letting the individual explore what is unknown, and even threatening. That is what will be educated by the games, and makes the individual autonomous within the variations compatible with the instincts.

The will offers new opportunities to those who can use the presence of the will to make it do what one would not have done when acting from instinctual patterns, as we shall see when we consider domestication of some species. Intelligence, a property already developed in many animals, and will working together will be the next domain of evolution trying to produce the new. Sometimes they will be at work in behaviors, creating new instincts, sometimes on the environment, utilizing its content to extend the power of existing behaviors to the point of changing them. An example of work on the environment is using hanging branches to pass swiftly from one tree to another, making it easier to escape some dangers as well as to get food. Sometimes will and intelligence will work on both behaviors and the environment at once. The attributes of the new species which emerge will be viable if the powers discovered are used in a manner compatible with the dynamics of the forces already at work in the universe. That "uncommitted brain" which is an attribute of higher animal forms is put to work. If it is uncommitted at birth that means that it is the first tissue which is available for jobs that have not been forecast in the sense of the unused sense organs before birth that can be put to use soon after birth.

In so far as we are looking at the freeing from instincts we see that the freedom sought for is connected with the instincts; in a way is conditioned by them. That is, that the animals we are considering having given themselves all that is needed to be helped by what evolution has done until then, do not give up any of these benefits and that their generation of the new begins from that platform. Generation after generation more and more of the energy made available goes into the exploration of what surrounds the individual and has made an impression. And this can be many things, mostly unforeseen. This develops a way of looking at the world that is not translatable into the vital or the molecular. Thus a wildcat close to a curled up porcupine will develop a behavior that is not of the hungry hunter unable to reach the flesh that will nourish it, but of the imaginative observer who forecasts that the porcupine cannot remain forever enclosed and will sometime open itself up again to offer its tender belly to the cat's sharp claws. So the wildcat wills itself to be motionless holding its breath and concentrating totally on the future moment that it can imagine must come, perhaps soon. It then strikes like lightning at the exposed and vulnerable belly and attempts to wound its prey fatally. Such a scene cannot be forecast before birth. To make such behavior available to an individual, evolution must have worked on freeing energy for the will which makes them possible.

Animal Mental Dynamics

The instinct is at work in the choice of the nourishment, making some animals prey and others not, living animals food and dead ones not. But we must invoke a more discriminating component in the "mental makeup" of animals if we want to understand what we observe in their behaviors. The idea of conditioned reflexes won't do because that can only explain that at a particular moment one perceives something and releases an unrelated functioning. Thus perceiving the curled up porcupine may trigger the reflex of leaving it alone or of kicking it to make it roll and find out if all in it is made of quills, or even of using a stone to approach the threatening needles. But it cannot suggest that one organizes the many components of finding a distance to place oneself so that the paw can reach the target in the shortest possible time, knowing that one's breath may be perceived by the victim and held for as long as one can, willing it so, to make oneself motionless by extreme concentration on the controlling nervous system. In a few words, in being as alert as possible with no expenditure of energy until the outburst. The contact of the individual with its energy must be such that it knows directly what its dynamics is and have it all potential and none kinetic, ready to reverse this order and make into kinetic energy as much as is necessary of that potential energy.

Knowing energy directly is the privilege of animals. Each instinct can be seen as one form of that knowledge. But as we have seen in the example above it is necessary to assume a capacity to transcend the instincts to account for such facts of behavior. It may be helpful to assume that "mental makeups" exist as soon as a form can know itself as identical with its behaviors. That as part of such makeups there are two functions, one concerned with the past and one concerned with the immediate present and future. In animals (as we did for humans elsewhere) we can call <u>psyche</u> the energy of the instinct present in the form and its functionings, and <u>affectivity</u> that uncommitted energy belonging to the individual and at its discretion to meet the demands of its projects.

The instinct makes the cat a cat and not another animal. Its psyche makes it at once know the significance of the muscles in the chest and the paws, the claws in the paws, the energy at its disposal. Affectivity mobilizes the energy in the creation of this existential response to the situation, collaborating with the psyche to have all of past experience at one's disposal, bowing to the instinct in accepting without doubt that the prey is nourishment and worth a fight, a pursuit which will enable it to endure as a living thing.

Animals which sleep use their time of sleep to be with their instinct. Their inner life is proved to exist by the fact of sleep. Indeed the fact of having to find one's nourishment makes animals use energy in moving in the environment which is known to be distinct from oneself and not accessible all at once or to the point that there will be no surprises from it and therefore moving in the environment produces impacts to be worked on. In the waking state their impacts from outside must be received and interpreted, and the instinct of the animal is involved in deciding which are to be ignored and which stressed. The stored up experience, the content of the psyche, is of two kinds; one is formed of the integrated impacts of which the instinct has already made sense; and the second is formed of the newest impacts including the fleeting, the strange, the exceptional. By giving itself sleep the instinct answers how to meet the new impacts to make them, when possible, into what can be integrated and made part of one's psyche when one wakes up.

The instincts therefore act is filters at the level of perception and what each animal takes to its sleep is conditioned by the functionings which through the instinct make one sensitive to some aspects of the world. But in sleep too the instinct will make sense of the retained material, sorting out, eliminating what is useless and integrating what is compatible and congenial. Once awake the animal will be a renewed individual not only from rest (if we know what that means) but because this recent energy distribution of the instinct upon the impacts of the day before, has integrated the experience.

It is therefore in the amount and quality of the experience one takes to one's sleep that we find another of the levers of evolution in the animal kingdom. Individuals become more so by adding a greater number of distinctive features that are more subtle and perhaps more effective than the fact of individuality resulting from somatic separations.

Discrimination and the Animal "I"

Individuals, because of their mobility can find themselves in circumstances which are new and challenging. Through accidents (of falling, being carried a-way in a current) they have opportunities to experience how new stresses on functionings existing in them can alter their responses to the demands of life. Adaptation to new circumstances is recognized as possible simply because it has been forced on oneself. We have found we needed the notion of instinct to account for perpetuation of sets of behaviors, now we are forced to endow each individual with a discriminatory power associated with the totality of its experience with the capacity to maintain the unique makeup and the unique retention of what has happened to on self. It is an energy center that commands the psyche and triggers affectivity as well as action, the one that changes each separate aggregate of flesh into an autonomous individual. In man we called this energy center "the self," which also changes each individual into a person. We need to acknowledge this in animals as well, simply because it exists and is at work, more or less, in all individuals. Let us call this uncommitted energy capable of making use of the uncommitted brain as well as of the psyche of each individual, the animal's "I." The seat of evolution is for every individual its "I" and, for the collection of the living, the way the various I's manage to coordinate, to cooperate , to compete, to fight with each other.

The relationship of the I and the psyche in each individual will serve as the bridge between what each individual does with heredity and what it does with itself in the environment. It is the seat of the will, of liberty in conjunction with the instinct provided. It is the I that tells an animal to leave a possible prey alone, to engage in a pursuit, to invent tricks to get its prey, to cooperate with others in manners adapted to the circumstances. The I is not a product of any of the functions of the instinct or the flesh, it is a product of evolution. It happens first because it is possible, second because it finds a place for itself that was not occupied, and hence not claimed, by anything else.

It is possible to generate an I because not all the time of the life of an animal is used for the struggle for survival. This last job is for the instinct produced with the form, by the inherited processes that permit the individual to make use of the environment. The available time beyond that taken up by the instinctual functionings may not be sufficient for some individuals to explore their opportunities, while for others, luckier ones, it may be plentiful. These individuals produce evolution, and their individual explorations which must be compatible with the forces in the form that permit survival, may be of a behavior whose consequences can be tremendous.

A New Kind of Being

Two of these explorations will prove to give individuals an escape from the instinct. They are the upright posture and the uses of the thumb. The second is more important than the first for evolution, but both make possible the creation of a new kind of being, no longer a mutation.

Looking at man as an animal has few advantages. But looking at man as escaping the third realm and establishing the fourth will be enormously advantageous in our study of the history of life on earth.

When the I of one particular great ape concentrates during a chance encounter of its self engaged in doing something, to notice that that individual's spine is such that it can be stretched, the same mechanisms that had been at work until then will serve to entertain the stretching. Only one individual's endowments, having gone beyond that of the species, is enough to reveal the advantages of the upright position and for that behavior to be carried on. Since it is compatible with survival, with the somatic functioning of that animal to give itself to knowing better what it means to be upright, it can do it. Since play exists, since relationship with other members of the species exist, it is possible that others can be struck by what they see and use their I to attempt to emulate the behavior they find attractive. Those who discover they cannot do it will abandon the attempt, keeping the phylum going. But those who through exercise can do it as easily as the initiator will provide followers and possibly a new species. The test of viability will follow as it always has in mutations.

Because of the contact of the I with the instinct, the instinct with the soma, the molecular-cellular components can be affected and an upright animal can result from this evolution. The DNA can carry the commands so that the vertebrae can be affected in such a way that if the animal is attracted by this form its posture will be upright. For weeks or months after birth the offspring will not use their I or their heredity to give themselves the newly acquired posture. They will possibly see upright other beings around and be inspired by this to engage in the exercises which will make them into members of the new species.

To be upright is only one of the features of these animals who have to test each of their individual behaviors for survival. Some will perish, others will need to do more with themselves so as to cope with the challenges resulting from dedication to the learned activity.

How many individuals try and how many succeed is not easily guessed at but it can be surmised that many at the beginning perished since concentration of one's mental energy on the new behavior may be at the expense of the concentration on the survival activities. For generations a learned behavior may not become somatically hereditary. It must be seen as possible to be learned by others who are attracted to it and are equipped for it. This necessary condition will make it the object of attention by other I's until it has been explored sufficiently for adherence to it to become collectively attractive. Then the existing mechanisms can take care of passing it on, via heredity. The work on the thumb can be thought of as going on at the same time, or not. It may have made a difference in the survival of the first animals that managed to become upright.

The fact that individuals have I's and can perceive in themselves certain qualities and turn their attention to them makes birth conditions part of the process of change. If someone is born with a deformity that makes the thumb capable of a different mobility than that of most members in the group, some of the behaviors of the species may be excluded from the constellation available to the others. Individuality permits this subject to explore what is open to it because of this "trouble," to concentrate on the possible, and to be struck by some advantage in the new condition. This may be cultivated by practice, refined and may lead to recognition of further advantages to that individual. A "defect" has become an asset.

Because not all the time of one's life is used for survival, it becomes possible for the young of the species – when bones are cartilages and their energy is at work deliberately in the muscle tone through games – to be invited to take part with that individual with the unusual thumb in exercises that extend the functionings of the hand. Those which can be extended in time beyond the ossification period, will remain as part of the constellation of behaviors of some individuals in the group. Their offspring will find in their midst these behaviors to emulate as they do others and can use the proven ways of affecting the next generation status within the group. The phenomenon of branching will have taken place and a new species will have emerged on earth. No doubt in some cases competition will affect this process of branching and some species may be unable to survive. If the initial "deformity" when dwelled in produces undeniable advantages, then the descendants will occupy the ground of the ancestors and their less competent cousins will have to be pushed out. The flexibility of the thumb leads to a pinpointed advantage: a different grasp on things. This advantage may lead to a capacity for example to detach chips of various shapes from rocks and to extend the function of claws already available through the DNA.

It is easy to see that 1) individuality, 2) time to play, 3) a better grip are sufficient to take the members of a group of apes from one constellation of behaviors to another and to produce a group that will generate a different approach to the environment where individuality, time to play, greater experience with the hands can bring offspring to a "weltanschaung" unrelated to that of the apes. Some of the changes can now be passed on to the molecular-cellular levels and a new DNA provides the individuals of the new species with means of saving time from generation to generation.

The Uncommitted Brain

It so happened that in a small group of individuals their explorations had a cumulative effect and that in a few million years to this date they have managed to have a significant number of offspring capable of adapting via the two processes of transmission and education, to a vision of the environment as a habitat to be changed.

As more of the behaviors were to be acquired after birth, the part of the brain which did not take care of ways of adaptation and of use of the soma was objectified to permit the entry and dwelling in the new environment. The "uncommitted brain" grew from generation to generation. Because it did not need to be ready to function at birth as the committed brain did, it could be a developing organ as our limbs are. The process of having potential brain available for years of life to come (but that may never come) is in agreement with the potential existence of the future. It is the future that guides evolution and guided it from the start. It is the property of uncommittedness of the brain which is compatible with evolution, not its bulk. Since it is found in so many species it tells us that evolution exists because individuals direct it. Its form shows that behaviors command the direction of biological change, that behaviors are more central to the process of evolution than heredity as transmitted by the molecular-cellular line which only provides the ground and the past. Since evolution is change, we cannot only look at the transmission of characteristics, however attractive we find them and fascinating the process.

The first "men" like the present "men" have their ties to the cosmos, to the organisms that made them and still make them part of "nature." It is still possible to see man as an animal. But is it helpful to us in understanding his passage on earth? Does it not generate pseudo-problems like that of the missing link?

If evolution is the appearance of the new on earth, we need to understand what produces the new. We know how perpetuation is taken care of at the cellular-molecular level. Why should acquired behaviors all be transmitted through that process? Clearly it will be most efficient for those behaviors that can be transmitted through heredity to be transmitted that way since it frees energy for individual use. But since animals do more than reproduce and survive, is there not an obligation to account for the whole of the reality rather than restrict oneself to one aspect and fool oneself that it is the whole?

Education in Evolution

Some men see evolution as an attribute of reality and state that everything, everyone in the universe, participate in it. Various people have used various lightings to help them understand the content of that intuition. The ones used in this writing have made it possible to integrate the cosmic, the cellular, the behavioral, and forced us to see that beyond the instincts that maintain the behaviors associated with the form from one generation to the next, there are individual I's. The emergence of the I's in individuals is required to account for the appearance of evolution on earth. But this to continue requires the passing along from individuals to groups of what has been gained by the explorations of what is compatible with the inheritance and the environments by these I's. For almost a century only the second challenge has been a puzzle for the people who were constantly studying evolution. Every proposal was found wanting. When molecular biologists understood the mechanism of the transmission of features via the structure of the DNA, it was believed that the problems would disappear. Today that problematic is as much with us as it has always been. This is mainly because only one realm was selected as the arena for evolution. Since we are of the four realms (molecular, cellular, behavioral and spiritual) we have to develop an approach that includes them and also include the future in evolution, since such a perspective is truer to reality.

***

In the approach presented here we have reached the point at which we see the effects of evolution shifting to the individual and to what the I of each individual can contribute. The function of the past is to free energy towards a more conscious universe. This process is the one we have met all through this study. We saw it in cosmic evolution that led to an impasse first at the atomic level and then at the molecular level, in the vegetable evolution which reached a different kind of complexity that tested the various environments on earth, taking each innovation to the point where it could no longer evolve. When in the animal kingdom we found the individuals feeding upon existing organisms, we saw a need for limitation of the number of members of species because of a finite food supply and observed the way this may force the transfer of evolution from the species to the individual. We saw that the individual changes had to be passed on to others and that this new challenge was met by a kind of education. The test of validity for this education was whether it allowed individual and group to become more compatible with the instincts and the environment.

Evolution that expressed itself in the animal kingdom by the trial and testing of behaviors has presented quite early attempts at regimenting the individuals as in the case of ants and bees for example. But even this did not completely eliminate the individual's initiative. This form of making the instinct both biological and social was abandoned and when herds, flocks, schools appeared, they did not represent an association with specialized functions for various individuals as in the case of ants and bees.

Recent mammals have chosen to live in small families sometimes for long periods sometimes for periods conditioned by the growth of the offspring. Generally individuals are sent to fare for themselves in the environment as soon as parents are sure that they have learned enough to succeed on their own. That education is their responsibility and cannot not be stored in the genes. The individuals have to decide whether their offspring are ready to be sent into the open world. Their instinct includes the behavior of becoming free of their attachment to their offspring once they leave them for good, but instinct cannot forecast whether they are ready to leave. Some parents push their offspring away and refuse to relate to them after a certain date and would not feel their link to them if they encountered them. The instinct in this case includes the behavior of dissolving a memory and of meeting that which was familiar up to one time and replacing it by a perception that it is new after that time. Bears do this. In other cases (gorillas for instance) the group may include three or four generations functioning as a family.

What is instinct in these cases is being tried out with these species but not necessarily adopted by other species which are then considered variations of these. That means that it has not been used by the molecular-cellular process to become transmissible by the DNA. It has been left to be adhered to or abandoned by individuals in their education.

From I to Self

When we look at man's evolution we see three things. One, the steady reduction of the instincts and the steady increase of the I's work; two, the emergence of an awareness of how to become freer by preferring to use the virtual to the actual and indeed using it, more and more; three, the increased and systematic use of education to compensate for the diminution of the role of the instinct so that today education is dominant and instinct marginal.

The first has made mankind a loose group of individuals directing their destiny. The second has produced civilizations and cultures, all the arts and the sciences. The third has made possible the grasp of what may direct evolution towards deliberate ends not formulated a priori but accessible to some through intuition and interpreted by them so that they seem acceptable to all.

Awareness of the Will

Awareness of the will is required to enter as grown-up into the functionings that need to be changed. Once it has happened to one individual, who may fast instead of moving towards food when hungry, it may appear as a gift of that individual or as a neglected attribute of everyone, part of the given. We can know the will better by putting our attention on easily noticeable functionings like delaying evacuation, holding one's breath, resisting falling asleep. We can know it even better by engaging in exercises steadily, studiously and noting their cumulative effect.

The consequence of awareness of the will is that the motor of change has been perceived, not only the change itself. It is the will that makes us pursue exercises that we would ordinarily shrink from because of pain. Thus it is possible to will one's soma into physical behaviors not necessarily smoothly compatible with the inherited expression of one's inheritance. There is nothing modern in the feats of acrobats, besides some equipment! The same motives that make some people today concentrate on extending their physical prowess can be assumed to exist in our ancestors. They chose to look at their behaviors and exercise their will to make one of them more effective or far-reaching.

Often in a puritanical society we can see girls at puberty stoop over to hide their growing breasts. In doing this they show us the work of their will which gives their soma a posture that is not inherited. Can we not assume that the adoption of the vertical upright posture could be caused by the same mechanism if not the same motive? It is necessary that the individual finds that the soma is adaptable to this; for example it is helpful that the foramen magnum (the hole where the vertebral column enters the skull) is (by accident or otherwise) under the skull instead of being behind it as it is in four-footed mammals.

Exercises are sufficient to make us adopt some posture and this assumes 1) a perception of what can be done so that it can be attempted and 2) a will to enter the exercise and to go on doing it in spite of the inconveniences.

For example, once we find ourselves jumping into water we can easily climb higher on the shore and jump from up there if circumstances are favorable. This can take us to considerable heights and our performances may be inspiring for others. Self-education leads to education of others whose will is directed similarly. Since the will is an attribute of individuals it is possible to find variations of its use that lead to performances that cover a spectrum.

Education

Animals too, can will themselves to engage in new behaviors, even if they have not made this a way of life as we find in stories in Ancient Greece, Sadus in India, tramps in the West. The fact that animals face difficulties may make us see that we do not need to consider a change of behavior the result of a mutation at the level of genes. Found by accident, or by an understanding of oneself brought by an opportunity, a new behavior may reveal its advantages and be adopted by the discoverer. Education may follow for others if they comply and so make them discover their capacity to display a behavior not transmitted by the genes.

When we include education as one of the means available to evolution to add to the inventory of behaviors, we generate a new level of functioning. Mankind will detach itself from the animal kingdom simply because it adopts education as a lever to transform itself and its field of activity.

It is possible to live the act of knowing in a variety of ways at different ages. Although the literature may not provide us with documents that prove conclusively that it has always been the case, we shall state our reasons for considering that knowing must have been present in the first men as it is with us.

First, we do not deny to bacteria the character of being plants because there are millions of more complex species; nor to amoebae that they are animals because there are millions of animal species that are much more complex.

Second, we do not deny that life is present in deciduous trees when they have stopped showing their vegetative life, nor in animals who hibernate or even sleep.

Third, we do not ask of any organism that it perform a certain function all the time in order to assert that it functions in that way. Mating, nest making, migrating are not continuous activities.

From these three arguments it is clear there is a variety of manifestations of any individual over a span of time. This will not be different when we consider thinking. It does not have to be as profound as Kant's to be called that. It does not have to be at the same level as the clearest all through a thinker's life. It does not have to be seen at work to assume that it might exist.

Knowing is an attribute of every energy present in any of its objectifications. We endowed an atom with knowledge of itself and of what it feels affinity for, and can relate to. Knowing of course has a spectrum for its expression and we need to keep this in mind. That knowing be purely human will deny the evidence humans have that in some respects plants and animals know some things better than them. Being closer to their cosmic origins makes them sensitive to what happens in the cosmos. Hence some animals have shown that they are alerted before people are that an earthquake or a storm is coming. People can regain sensitivities which plants and animals display.

They need to <u>re-educate</u> themselves in order to do it, re-educate by returning to their origins: the other three realms.

But in the human realm there is something new to become vulnerable to and we call it awareness.

Human Evolution is the Story of Awareness

In the development of the human species, awareness moves from appearing in the beginning as a possibility compatible with the instincts to becoming more and more itself, integrating of the past all that which does not hamper its working and shedding away what does.

When those animals that stressed the will and more particularly the will as it can manifest itself in the individual, started to take it seriously and let it guide their actions, they opened the door to the escape from the instincts. A large chunk of evolution for many generations must have been taken by the job of knowing what can be done with the will by proposing to it many challenges that only it could meet.

There was no dearth of challenges that by instinct one would shrink away from but that the will let one take on, at first hesitatingly and then less so as one became acquainted with them until one knew enough to consider them one's own.

Swimming is one of them. Man had to find in himself the will to enter the water, to discover that floating was possible. Perhaps in an accident the movements of his limbs in his fear made him know that it was possible to float and propel himself and try it later deliberately. By selecting shallow waters he could experiment with his movements, invite his young to do likewise and integrate swimming as a human behavior without having to imitate another animal. Not all humans had to learn to swim and many had a full life without ever entering the water. But swimming tells us that new behaviors can be added without essentially changing the nature of the rest of one's life, without having to be aggregated to the genes as a hereditary feature of man. The existence of the will makes man capable of being different simply by finding new uses for the old equipment.

The development of techniques of swimming will be undertaken by those people who, having to concentrate on it and having the time to devote to it, will discover what hampers and what helps. It is clear that only oneself and favorable circumstances are needed to improve on that behavior. But it is also clear that the individuals involved must concentrate on the activity and become aware of what they are going through. The inner movements are the <u>sine qua non</u> conditions for the study of such activities. Two new attributes of man emerge: peace at making mistakes and persistence until the obstacles are dominated. Animals already know both but only within their instinct. Mistakes are measured with the yardstick of the perfect behaviors that their instinct recognizes. They make mistakes because they may be fooled or are daring, but man seems to know that he is encountering the unknown and cannot meet it without making room for it. Since he does not know what it is, he includes as a component of the situation that he may guess wrong and needs to try again and again. In between, the behaviors that are wrong must teach him something valuable and lasting. To learn to see mistakes as normal forms part of the evolution of man, of his awareness. To test, to question (not necessarily verbally) is the way the presence of awareness is made manifest. Built-in awareness is experimentation. To allow oneself to be provoked by one's sight; by one's touch, by one's sensitivity to sound, one's inner sensitivity, is equivalent to acknowledging the existence of awareness.

Already foreshadowed in the previous three realms awareness comes to its own in man, not necessarily, inevitably, but as a matter of fact. Some individual having energy to spare makes the observation that it is observing something in which its "I" is involved and retains some of the features encountered in the observation. This retention prolongs the life of the observation and gives it a chance to be integrated into the individual's life so that it is capable of being triggered again. Thus, the activity of turning stones to see if there is anything under them starts with some individual doing it and finding it occasionally profitable, and so worthy of being maintained. Its "I" has asked a question and instead of an answer that cancels the question found that the question remains, becomes a tool in its exploration of its life in the environment. It could be passed on to its offspring and relatives because it is beneficial. It becomes an acquired behavior via education for the members of that group.

The Need to Know

In man such attempts can be seen as the rule rather than the exception. Man is characterized by his need to know. To know has become his main preoccupation simply because it was possible and he entered it, not only for feeding and surviving, but to find out.

He will engage wholeheartedly in any one thing that strikes him; and devote himself for as long as is needed to knowing it. And there is so much that can strike him. In the beginning he will be attracted by what requires himself alone and his presence in the environment. Thus he may find that his ear on some ground perceives the sounds of some distant movement, simply because he is stretched on the ground and his ear is against the ground. He may devote himself not to understanding why, but to knowing the features of this method of surveying the distant environment with respect to the noises in it. He will thus know how attention and alertness amplify the energy received and he will use it as an instrument to interpret the messages, developing an experimental method of his own, matching what he concluded he had perceived with his ears to what he can find out using his other senses, correlating them and retaining the connections when they are verified and found useful. In such a way he may endow himself and others with means of knowing not yet part of the group's equipment and it can stop there. Surveying up to where the boundaries of the validity of the instrument are may take a long time and absorb many lives. Considering the large number of phenomena that may strike one, it is easily understood that the evolution of awareness has taken a long time. Millennia are required before a sufficient store of experience that can be re-discovered by each new individual showed a marked difference in the quality of life in some groups. Still simply to become aware is enough to start a new trend and to remain engaged in finding out what it can contribute to one's life and through this, to other lives.

The history of awareness is not identical with the history of inventions. Inventions are organically connected with awarenesses but not the converse. The invention of fire, the wheel, farming are momentous in man's history. They are mingled with a number of observations and some daring actions, and resulted in being instruments that made man more adapted to the possibilities on earth. But they are not as important in the long run in the evolution of mankind, as the awareness of what imagery, intelligence, or the self, can do on behalf of each individual. Indeed with fire, man can destroy man or nature but intelligence may show that this does not need to be pursued. Without the wheel the Incas had a remarkable civilization and with farming men could only feed so many until intelligence produced chemistry and fertilizers and altered the dependence on the content of natural soils.

Evolution in Man is Molded by Awareness

We have many civilizations and in each many cultures and in each of these various societies, communities, clubs etc., because man has reached awareness of his singularity within his individuality.

The sacred books of the various religions tell us the story in a variety of ways. They are not only depositories of how the generation of the writers received the religions and transmitted them but also the collection of the vision of evolution of those involved, the collective wisdom or the gathered experience of those most aware among them.

All the religions have been needed to tell the journeys of awareness. All the cultures within them represent syntheses of the awareness of the founders with what is left of the three realms plus the constraints and opportunities of the environments plus what some individuals did with themselves in the given circumstances.

Some individual man whose movement of awareness was such that it could appear to many others as real and therefore as acceptable, could influence the movement of awareness of others and polarize their behaviors towards certain forms and constellations. It is a remarkable fact that for millennia, spontaneously, human groups have held the words and lives of some people as the guiding lights of their own lives. Around 2500 years ago a number of men spoke and their words are still heard repeated with respect and have been loved by more than one hundred successive generations: Socrates, Gautama the Buddha, Confucius, Jesus (a little more recently). Before them Abraham, Moses, Rama, Krishna, and after them Mohammed and lesser leaders up and down the planet, have offered to their contemporaries what their awareness had reached. However much the challenges of everyday absorbed the lives of the generations which came after these men and women, they also found in themselves the echoes of the awareness of those who proposed renewals of awareness.

This particular phenomenon is important for us in two respects. First, it tells us how essential is the role of the individual in evolution and second how a universe of experience opens up when one awareness is compatible with all that mankind has done with itself in the past, but is capable of recasting these previous evolutions to renew the human universe.

On three occasions in the cosmos and on earth, it has been possible to integrate the existing into the new.

The first was when nuclear aggregation had produced all the possible atoms in the cosmos but could not maintain the vector and had to start a new development and, under the guise of molecular reactions, produced another layer of duration and the world we now know. The second was when also with still smaller amounts of energy the vital started, on top of the cosmic, its vegetable evolution which depends on the possibilities of synthesis. The third was when with still less energy in each transaction the animal kingdom showed what can be done by analysis of the cosmic arrangement in the previous two realms to produce the evolution of freedom on earth. Freedom of movement, freedom of choosing what to integrate and what to leave alone, and the successive freedoms which may viably be associated with some constellations of behaviors.

Freedom Explored

Now some men have found that there are other universes for the individual to explore and objectivate with as little energy as possible, by working on one's awareness in specific ways. They made education the process of passing on to individuals what the group had accepted as the expression of its exploration over the generations. In a few of these civilizations, awareness became the item to work on. In that realm self-education replaced education and the visibility of the energy used in that transformation disappeared to the point that there are today people – whose job is to be concerned with energy-who deny its presence in the phenomena under scrutiny here. The whole movement towards the use of less and less energy to achieve greater and greater transformations, has not yet been perceived by most observers. Yet only from it does freedom result, freedom from what exists already and has been objectified so that new manifestations can make more explicit the presence of energy working upon its objectivations.

It is possible to be lost in the details and not to manage an overall view of the whole. It is also possible not to be lost in the details and to be lost in the view of the whole. It may be possible to remain in contact with the details via the whole and conversely. This latter way of working deserves more attention. What has been preserved from the old civilizations tells us only something of what occupied the men who lived in them. Much has been lost and can only be guessed at through consistency and compatibility with what is known now. Some of the old civilizations are still alive and thriving. They bring to us documents which serve to inform us but also to challenge our conceptions of man-in-the-world. Rather than consider ancient ways of knowing as primitive and the forerunners of our own ways, we can expand our sensitivities to make the original ones mean to us what they meant to their originators. When the positivists gave themselves the place at the top of the pyramid of knowings while giving mystics and mythological thinkers a much lower level of manifestation, they simply were arrogant. They served evolution by giving us the explicitation of their grasp of the world but they made the mistake of believing that they understood something when they explained it away. To widen our ways of knowing to include the ways of the mystics and of the mythologists it is to be true to the actuality of an evolution that still has a place for mystics and is uplifted by the symbolic significance of myths. We needed to grow beyond positivism to be able to integrate into our ways of knowing these vastly more complex ways in which the self engages in dialogues with as much as can be grasped of what is.

Evolution in the Fourth Realm

Looking at the various civilizations as manifestations of the alternative evolutions open to man and looking at them together as the ways men use their time of living on earth would we find a new meaning to evolution in the fourth realm?

Indeed we need to strain a little to understand some of what we find in various civilizations in terms of the other civilizations. There is a profound difference between being deeply immersed in one civilization and living only on its terms, and being made to contemplate such a life in terms that do not belong to it. This difference may be sought as the acid test of any synthetic vision attempting to find means to reconcile what has evolved so differently. Acceptance by the "orthodox" of a relativistic view seems indeed the only criterion of the realization of the proposed synthesis. In a simultaneous study of men immersed, as mankind is today in a variety of cultures, religions, languages, we shall know that we managed to reach the true springs of mankind if each phenomenon appears humanly possible conditioned only by earthian circumstances and the fact that it takes time to live experience.

Libraries are full of records of observations of rites in myriads of communities. Films have recorded the outer appearances of tribal ways of living. Students over the years have produced comparative studies of art and music, dance, pottery, contrasting methods of production, tracing influences, making explicit what touches them as observers. In each culture or community there is a story of evolution. Not all civilizations produced the Mahabharata or the Old Testament which still inspire men and women by the millions. But all civilizations operated by handing down to those who were born in them a body of doctrine that spoke straight to their gifts and was compatible with all the ways of knowing available to them.

Because intuition is the most comprehensive and therefore the most primitive of all ways of knowing it is found at the basis of all religions and all civilizations. Since all men start by being children uncommitted to anything but acknowledging what is and what comes their way, in the beginning of the meeting of awareness of oneself as a knower we find that there is no doubt, no philosophizing, only the assertion that something is and the experience concerned with it. Reality and perception are considered as one. Because our sun is mainly beneficial and its energy is at the source of so much that involves man, man will relate to the sun, infinitely more powerful than him, in a relation of dependence and of respect, possibly of love. Because the moon looks cool and gives us little except in the mystery of its cycles and the softness of its light, man may be inclined towards a sentimental rather than vital relationship with the moon. Storms, earthquakes, the forces of nature, as well as the variety of the vegetation and fauna on earth appear unpredictable, threatening, a transcendental given, to him who is powerless to do more than acknowledge their existence. It is easy to see how this perception causes man to lump together disparate items moved by mysterious forces. Closer to home is one's inner life. The dialogue of perception with the world around is not of another nature than that with the world within. Like the baby who cannot doubt the energy that impinges on him and makes him process it from the start without labeling but not without full consciousness, first men did not have the problem of a dual universe of mind and matter. All was in all, everything was inter-related even though not so many questions assailed them as do us today.

Mythologies

Mythologies are organizations at the level of the complex, of intuition, of all that which man's awareness has been able to perceive and consciously acknowledge. They speak of the penetration of totality by minds using all they have, all they are, to make sense of the universe of perception, action, feeling and thought. They use images, metaphors, allusions, to indicate that they know the intimate links between language and experience, language and truth. They also tell in the form of stories that can reach the hearer, how affectivity, intellect, will, are perceived as at work sometimes together, sometimes in their contrasts. Myths also reflect the accumulated wisdom of the generations which is nothing other than the capacity to learn from life, to acknowledge the pitfalls that lead to common errors, errors recognized as an integral part of learning to live.

In mythologies, people store not only the substance of what has mattered to the more advanced members of the community but also the subtle educational levers that have been found successful. Stories can be heard again and again and related to at different levels, because they use materials and procedures which have proved over time to be capable of affecting the minds in the community. Sometimes this is because of the vividness of the images as triggered by words, sometimes because of the intellectual horizon lit up by the example. The language, itself chiseled as if it were stone, through its sounds and rhythms enters into the sources of mental energy to stir more than ordinary speech can do. Because of this people tell them over and over, they become familiar and so their content and educational value is more easily available to the community.

Mythologies serve a number of purposes in the education of the awareness of the members of the community at large.

1   They hold what has proven valuable in nourishing the mind of the community. Called archetypes or something else, these items had been deliberately brought together, in some perhaps remote and forgotten past of that community, because someone found they had value for that one first and for others subsequently. They remained as part of folklore because they were found to be capable of generating some echo in the minds of the hearers. Their selection for retransmission, at least in the beginning, could not be ritualistic as when some stories are told on certain occasions or festivals. They must have been retold because of the effect they had on the people who were the implicit targets for them. Their liking them is the test. Somehow the alchemy of the mind will get hold of them and transmute them into lights for action in situations where doubt presents itself. The usefulness of the stories assures their perenniality.

2   Myths testify to the level of evolution in self knowledge of some of the members of the community. This knowledge of oneself, has perhaps never been verbalized, but it can certainly be experienced in the area of expression: verbal and pictorial. The authors of mythologies however anonymous, must have known the special philters to use when selecting words and rhythms to maximize effects. They must have had specific criteria for aesthetic performance as well, and this can show itself as much in the form of the story as in the fact that people ask to hear it again and again. On an absolute scale we might say they could reach peaks at once and stay at that level as generation followed generation trying to climb on the shoulders of the original writers or speakers but never managing to do better than them.

Evolution here does not mean progress. It refers to the passage of something valuable from one consciousness which reached it to other consciousnesses that might not have arrived at it spontaneously. Taken together, a mythology offers a universe of experience available to all as a banquet table where all tastes and all appetites are catered to and more. The "more" is concerned with self-knowledge resulting from measuring oneself against the characters of the stories adding one's particular personal chores and jobs to the inspiring ones. A mythology simplifies the world in one way but complicates it in another. Not all acts in life are taken into account, but some are represented as more important than others. The listener is forced to receive that awareness.

3   Collective wisdom results from awareness by one individual first of what seems to help that person cope with life, and then of

adoption by others of the relevance to their own condition of the experience mentioned. Mythologies are the repository of wisdom as well as a standard of beauty and of a common sense approach to challenges. They are not a primitive element of the organization of life, since they transmute experience into symbols applicable to many and therefore can only come about when much has been gathered, and a language developed. They indicate that a group has gone quite a long way in its climb on awareness.

The assimilation of mythologies is part of the evolution of a group through education. The contents of mythologies are those elements the group wants its members to consider as part of their psyche. They are the link for individuals between their own past and the past of the group. This past is not made of the recall of events but of the significance of living as it struck keen observers among the ancestors. Mythologies refer to events to produce stories but they transcend events in order to offer lessons to generations far removed from such events. Group evolution present in the offering of the mythologies becomes a lever of individual evolution because of the consciousness of the individual in the memory of the story. What the mythology offers may seem to remain one and the same because written in a language that becomes archaic and dates the events referred to. But what is there is a dialogue between awarenesses or consciousnesses on some specific point of awareness. Hence it is always recast by each individual sometimes several times. One assimilates a mythology and thus belongs to a group. This group exists only because of the dedication to the messages and the participation of the individuals. The more the participation, the more cohesive the group. Sometimes a group demands of its members total allegiance to the message in the mythology.

4   Mythologies are psychologically true even if they are not historically true. This means that the criterion of truth is their spontaneous ability to perpetuate their content from generation to generation as part of the individual's psyche, springing to consciousness at anytime as if called in by the perception of a universe through someone else's eyes.

Symbols are the flexible reality akin to the psychic stuff that fills the mental life of individuals of any generation. The truth behind the symbols that has reached through the people who offered them, is what recreates a reality compatible with the truth and with the circumstances of one's generation. Symbols per se could not influence men. It is the truth they make apparent that does. The makers of mythologies knew it when they chose among the stories current in their societies certain ones to save for the coming generations.

5   A complex grasp of the world with a logic compatible with fantasy and art, is the foundation of a mythology. This is true intuition and intuition of truth at the same time, since no writer knows the readers of the future or what their tastes will be. Because of that, readers or listeners do not demand realism from the storyteller. Inspiration is all they require.

Some of the storytellers' intuitions leave us panting and ready to concede supernatural powers to the heroes as well as to the narrators.

Mythologies belong to evolution because they free individuals by inspiring them to see the world as larger than their perception suggests. Once in that state, the individual, inspired to forge ahead and to pull the group behind him becomes the true agent of evolution. Mythologies at the same time belong to and transcend traditions.

Because of this they will always be important. The exact sciences cannot dislodge them from their position. They can only try to emulate the way they nourish the imagination, the symbolic powers of the mind. They can only seek to find the secret of their perennial attraction and success.

The way people or gods behave in myths reflect the ways of knowing available to the founders of religions. The heroes or gods become real to anyone who entertains them and that is what groups suggest to their youths. A reality based on the mythology is generated. It gets hold of minds and produces participation. That reality can supplant others in one's awareness and become intimately part of the fabric of Reality as one knows it. Some truth associated with the various realities gives Reality the right of claiming to be true, although it is relative because of its being a construction by minds at some stages in their own evolution.

When traditionalists hold to the letter of the expression of a mythology they reveal that they do not know what mythologies have been created for. If revolutionaries reject them as "opium of the people, " they do not realize the evolutionary significance of mythologies, the fact that they dwell in the past only symbolically through symbols capable of many different interpretations, thus allowing the people they inspire to take their own steps outside tradition. Mythologies like sciences function as clearing houses: they receive the input of some to be made available to others. What they make available is truth about the subtle substance of conscious living in a form called symbolic because it refers to rather than is, the substance. Because it is truth they propagate, they remain in circulation for hundreds and thousands of years. Because that truth works through symbols it can be recast again and again according to the changed styles of the generations but always recognized as significant to cope with the large and the small events of one's life.

Mythologies keep the poetic vein running because symbols trigger images which do not need to conform to anything, only carry the dynamics that ensure that one is living more fully in the moment.

The creators of mythologies know of ways of knowing and of presenting the knowledge reached through them that give them a very high rank among the seekers of truth. Instead of abstract statements about the unveiled truth. such as have been adopted by the modern sciences, they offer us words arranged in such a way as to illumine in us complex inner functionings, close to the real dynamics of the mind.

Since myths have a considerable power of inspiration but are creations of the mind, their creators hold such high positions in our esteem that they are generally labeled gods or goddesses.

They can be rejected by a narrowly defined reason adopting temporary political attitudes against their universal appeal, they still remain the channel given to truth about conscious living by our ancestors and an important lever of human evolution and human education. Everywhere on earth, in small valleys or large areas, myths have been what human beings found to be the form best suited to convey their psychological growth, the self awareness of some on behalf of all. Myths require the simultaneous mobilization of a multitude of centers of energy: some coming from neurons, some generating links between the neurons, some from the self to produce the new constellation called the impact of the myth. The self also brings consciousness in and from it results the unique individual memory of the myth, which will guide a unique meeting of life.

Far from being primitive uses of the self, myths represent an extremely able way of handling matters much more complex than those encountered in laboratories by modern scientists. As scientific facts they have to be seen as the facts of the science of myths, not of physics or biology. In the science of myths we find the facts of myths and they are as worthwhile as any other facts and as fascinating, because they were systematically studied thousands of years ago by methods we no longer practice. These methods are lost to us simply because we abandoned them, not because they were useless. In fact, now that we look at the whole of the evolution of mankind, we are finding that we are vastly helped when we can understand the ways of working of our ancestors rather than judge them with spite and move ahead without any serious examination.

Horizontal and Vertical Evolutions

The ways of working of our ancestors tell us that there are two kinds of evolutions in all the realms. On the one hand there is what we shall call "horizontal evolution" when we see groups exploring the consequences of one awareness or a number of inter-related awarenesses. On the other hand, there is "vertical evolution," which is a new way of handling energy. Most horizontal evolutions end up in impasses after extensive, intensive and extremely successful collective experimenting on, with, and around, the awareness or awarenesses. The renewal of mankind takes place when someone finds in himself and his life a way of handling energy that integrates – by transcending it – what was available. This vertical evolution has been at work as we saw a few times in the three realms of the cosmic, the vital, and the instinctual and many more times in the fourth realm of man. Today we are witnessing one more of these leaps.

***

Horizontal evolution is the stuff of histories. Vertical evolution is needed to produce the germs of new civilizations. The study of evolution covers both. Collective memory only retains what succeeds. Individual memories contribute little to collective memory since the collective must integrate the individual in denying its individuality. Collective memory that is carried only by individuals, has found its way of being passed on from generation to generation by being integrated into many forms of transmissions including DNA, rituals, what is allowed in nourishment and institutions, belonging to all four realms.

No wonder horizontal evolutions end up in impasses. They are the unfolding of one possibility and when all has been found, with this goes the definition of the end. A new beginning is needed. A new universe is required that may only be defined within the old, but with attributes that were not possible in the previous circumstances. Man's awareness of evolution of himself has made him see evolution everywhere. He sees in the cosmos the evolution of atoms using high levels of energy to generate the new atoms. This evolution changes the content of the cosmos. The cosmic laboratory in its manner finds all it can do and stops or rather goes on doing well what it can do well when consuming huge amounts of cosmic energy. The impasse occurs when no more atoms can be obtained in the cosmos because the horizontal evolution has produced all that can result from that process and with that energy. A jump above in a vertical evolution, "chemical energy" makes its appearance as a way of elaborating the cosmos but it appears only in certain places, i.e. those places where the energy of electrons suffices to produce the new molecules of the world. Because conditions permit the interchange of outer electrons between atoms, molecules make their appearance, transforming those places in the cosmos. A horizontal evolution commences that leads to an impasse in the cosmos when no new molecules can be formed because as soon as they are formed they break down. Still that horizontal evolution that has changed the world locally, has brought the world to the point where another leap vertically will show energy using itself more effectively in the cosmos.

Knowing what was successful in the cosmos, life starts from there to produce another horizontal evolution generating on earth, all the vegetable realm. This too leads to an impasse requiring another vertical leap creating the animal kingdom. There energy again operates on the existing, the successful, to produce the horizontal evolution of the instinctual realm. And another impasse.

The energy of the animal is at work on a keyboard that only commands variations of somatic energy as can be seen in the case of the muscle tone.

This new level of "differential use" of energy in order to produce all the behaviors in the universe will also lead to an impasse for the ensuing horizontal evolution, since this is still conditioned by the given which is the vital and the cosmic as well as the residual energy upon which the animal plays variations.

To proceed further, another vertical leap is required, one in which residual energy turns away from the forms to which it was so associated as to be identified with them, although it was different from them. Free energy has come to be and be acknowledged as existing before all of it is locked up in forms only linked by residual a-mounts. All that was successful in all the previous realms can now be integrated in a fourth realm, devoted to the exploration of what free energy by definition can do with itself. As easily aware of chemical reactions as molecules are, and of vital transactions as cells and tissues are, of behaviors as animals are, men, the beings endowed with self-awareness – as they call their level of energy capable of directing itself upon itself – men will begin an adventure sui generis. Aware of energy, men will consider energy variations as the touchstone of their activities. Practice in noticing as many variations as possible, will provide constantly more contact with the attributes of energy, as well as interest in the economics of energy involved in every activity. This leads to the recognition of energy wherever it can be found. When energy is seen in animals they will be domesticated; when it is seen in plants, there will follow farming, husbandry and cooking. It will be discovered in slopes and waterfalls, in stones and minerals and will lead to the exploration of gravity and tools. Each of these adventures is a horizontal evolution. Each will lead to an impasse coinciding with the discovery of the limitations of every awareness pushed as far out as could be.

As awarenesses do not necessarily involve a form as was required in matter, plants and animals, they can be known only to the individual who has them and they can be wasted for the group even if they could be of paramount importance for its evolution. Socrates among the Greeks knew the significance of being aware of awarenesses. In spite of Plato his group passed him by and the Greek civilization went on for centuries unaware that his expression was compatible with the rest of its traditions.

Two thousand six hundred years later the West is in a similar situation needing to make the leap represented by that new awareness and seeing only its traditions as salvation.

Every Civilization is Embarked Upon a Horizontal Evolution

Some have reached their impasses and we only know their objectified remains. Some are still flourishing and some are in the process of changing. Each is rich in all that the people in it have done out of the proposals of their founders – made elliptically and symbolically and needing centuries to be made explicit.

Is not horizontal evolution precisely the unfolding of the energy content that can be released from the nutshell of the message of the founder of a religion that becomes a civilization? The expression by a succession of generations of what has been revealed is the content of that civilization and at the same time the explicitation of the message. The energy in the message is at another level of potential than the living of individuals and the groups they form. There can still be equivalence between all the energy used by the civilization over centuries or millennia and the energy reached by one man in his awareness. From that higher potential, representing the leap, the new energy can lead other energies with lower potentials to attempt what is compatible with all they are to manifest some of what is implicit in the message. If induction is a model from electromagnetism, people are susceptible of being induced because they are energy submitted to an inducing energy capable of induction. The terms in the various fields of life may vary as they vary from language to language. But in all fields, to mobilize all energy dormant in every one only needs a sufficiently strong inductor. Founders of religions and of civilizations, seem to be that and to know how they can affect others.

Human evolution in a number of horizontal attempts called civilizations, starts with one person (Abraham, the Buddha, Jesus, Mohammed, among a number of founders) whose life and statements are testimony that one operates at a level of awareness such that it represents a vertical leap at the moment of living and the establishing of a new way of being possible for all followers. That there are followers proves the truth of the experience. Each civilization at the same time integrates what had succeeded in the previous way of living, generating a continuity for mankind.

Each civilization generates cultures which may differ from each other in so far as they also integrate conditions on earth, which may vary from valley to valley. Cultures like civilizations, can lead to impasses and vanish. In each, people live, take part, and must take responsibility for its success or difficulties. By refusing to experiment with what comes, to test its compatibility with the message of the founder and allow it its right of place, some people interfere with evolution, yet they believe they are being faithful and orthodox. By undertaking to experiment with what presents itself in relation to the message of the founder, people evolve. They must first take a leap vertically and reach a level of awareness which, they can confirm, agrees with the message, and from then on they can live horizontally unfolding in their own life what it means to be at that level of awareness.

But because different individuals are bringing to their new level what they were successfully till then, the collective expression as a relief of a multitude of different individual expressions blurs the fact that all belong to the same culture or civilization. Stressing commonality or ignoring differences enhances the sense of belonging to a civilization or a culture and gives each a reality that can be recognized by outsiders. In one culture or one civilization horizontal evolution cannot be in one plane. The thickness of a layer can account for the variety of individual living of the message, the parallelism of the planes in the layer represents the commonality. Still the variation caused by one individual may be such that the layer is not only widened but transcended. Depending upon whether the leap is acknowledged as compatible with the message of the founder or not the individual concerned will stay within the fold or begin a new evolution based on that leap. For example the Reformation was counted as a Christian leap though outside the Catholic Church, while Buddhism although compatible with Hinduism, is not counted as part of Hinduism by Buddhists. Buddhism is a vertical leap from Hinduism that produced its own civilization. Protestantism only produced cultural changes within Christianity.

The fact that so many religions co-exist on earth at this time, that many civilizations side by side share the planet, tells us that none is really the expression of the latest leap that may transcend them all. At this moment we see a number of horizontal evolutions still going strong, vaguely competing to absorb the planet.

The Creation of Language

A similar situation results from the co-existence of a few thousand languages on earth for the expression of elements of the various realms: perceptions, actions, feelings, thoughts, images, projects of various kinds. The leap exists that made men capable of finding the basis for a systematic examination of utterances as a means of expression and for communication. That leap which happened long ago must have been followed immediately by objectivation of an integrated system of perceptibly different sounds to correspond to the various perceptible properties of reality. Because the creators of language everywhere own awareness they could classify their perceptions, note the changes when light and distance affect things in their appearance. Reality was simultaneously variable and stable according to what one did with oneself. Stability resulted from memory and evocation of impressions comparable with present perceptions; variability resulted from mobility of either or both, the subject and the object, or from circumstances on earth. Therefore everyone on earth aware because yielding to impacts would know reality as a set of impressions appearing sometimes as different, sometimes as the same. The observer develops a sense of transformations as part of the perception of reality as acceptable as the working of one's sense of truth.

These awarenesses are essential and must have taken place before men could develop any system called a language. The basis for the creation of language is a set of awarenesses of inner changes affecting what one associates with outer changes. Everything is happening within individuals who must find it in themselves before taking the initiative of associating somewhat firmly one thing they do – utter a sound – which is accessible to others and another thing they do which is perceptible only to themselves, to their sensitivity.

Only beings who are aware simultaneously of their awareness of themselves and the functioning of others, of what goes on in them in the presence of others and in the presence of objects or processes or perceptible variables, can create a language requiring and reflecting all these awarenesses.

We must therefore conclude that all tribes on earth who have developed their own languages must have had prior to it access to their awareness, to its dynamics and to a distinct knowledge that their fellow tribesmen were functioning like them. In addition they had to be equipped with all that is necessary for the specific tasks of producing language and that is not little.

Indeed, languages everywhere display economic features. The corresponding awarenesses must be assumed to pre-exist clearly in some minds if not all. Part of this equipment results for instance from awareness of what one sees and uses.

Transformations as awareness come before the creation of languages and they have been incorporated into language from its beginnings and do not stand out or call one's attention. Stressing and ignoring also are needed for perception and educated by perceiving, from the start. Awareness that correspondences exist is also needed. With that equipment one single leap such as the use of one sign to refer to something else, may precipitate simultaneously the deliberate use of a number of functionings associated with one single signal.

What we are saying is that there are means in the mind of men endowed with awarenesses for the production not of one language but of as many as we want, for the way a correspondence is produced between the contents of one's mind including the dynamics which are real and true, and a set of arbitrary signs and rules, involves a choice which is entirely left to the creative individual. That is, the choice is left to someone who does not have to pay attention to any restriction other than accomplishing what he aims at: the correspondence, as a perceptible system, coherent and at hand. So one will utter a particular noise that can be distinguished from others. Tone and stress are as good ways of distinguishing between impacts as sounds. Hence we find some languages electing tone and having as many as six or seven to broaden the vocabulary, and others, not using tones, using deliberate algebraic transformations such as permutations of sounds to produce new words. Clearly all this is only concerned with one tiny feature of languages but it illustrates well 1) that we cannot avoid calling in awareness to understand the creation of languages 2) that once we know that awareness is needed we can study the uses of awareness by ancestors engaged in a very specific human endeavor, the creation of language, and know what they had to reach in their conscious evolution to manage such varied feats.

We as babies found in ourselves the same endowments and used them so well that we learned our first language very early and for good.

Man's Evolution

In man's evolution we can see two engagements of awareness, one leading to horizontal evolutions requiring the maintenance of one's awareness in contact with what is being lived, and one leading to a vertical evolution when awareness finds in itself a property not yet noted and enters that realm instead of going on with what one could do.

We called "property" what struck our awareness. There may not be a word for that awareness of one's awareness. Property will perhaps serve for a while if the reader is careful and supplies his. own examples to illumine the language used.

No doubt the co-existence of a somatic self, a psyche, an affectivity, the free energy of the self, awareness, the will, intelligence, as attributes of the self, perception, action, feeling, sensitivity, activity, functionings, all taken together cannot be sufficient to account adequately for all we can know. As we evolve we find new universes of experience, create new tools for their exploration, require extensions of our languages to talk of them. This is the case here when we want to contrast awareness of all the manifestations of life which necessarily involve in the process awareness of awareness but only for a fleeting moment and the permanent awareness of the awareness which is a leap that can generate the next layer of evolution for man.

When a being noted awareness for the first time, man was capable of stepping out of the 3rd realm and of beginning the 4th realm. Awareness noted did not remain the object of that man's awareness. Had it been, we would not have found on earth all we have found. Instead men applied their awareness to what they were and what they found in their activities, both to the satisfaction of their needs (of the other 3 realms) and to the satisfaction in understanding some of their actions because of the new light available. They could now turn that light upon every one of their manifestations and change their relationships to their environment as well as change the environment. Much much later they could summarize all their findings in their mythologies and their arts. They took the time to become aware of everything they felt with their hands, knowing the universe of touch. They studied how looking could serve seeing and listening, hearing. They found that one of the terms engaged their responsibility and the other a world they did not make, which existed beyond them, objectively, irrevocably, challenging them with its content and dynamics so as to be known, used, mastered.

The many dialogues that opened up between the self of each individual and what is not it, took the time of the lives and the passions of those who let themselves be absorbed by the challenges. Every generation made its contribution through some of its members without leaving documents or evidence but leaving behind know-hows and experiences that went to form the collective knowledge, the collective memory of the group. Education, then, is the outcome of the exercises, the games offered by those who knew how to those who did not know how yet. Integrating in one's self what the generations found essential for survival first, for progress second – was as always, self-education, the only true education! Those who only achieved this acquisition of the acquired progress joined the group to maintain the collectivity where its conscious work took it. Those who found openings through the exercises and pushed forward reaching greater performance in some areas pulled the group ahead in its horizontal evolution. Occasionally one or other would reach beyond what was possible with the transmitted powers and find a slot through which to soar in a different realm compatible with the original awareness but unveiling new possibilities for that awareness.

It is not a too difficult job to illustrate this process of work of awareness engaged in the world. For example a universe of being – lived through automatically until then – becomes suddenly a source of endless discoveries that affect the group irreversibly, revealing a world where one's awareness could be totally absorbed. Until such events come one's way it seems unbelievable that so much has been ignored for so long but now solicits one's attention as an absolute. Who can believe that all those who during decades in their lives used their intellect in matters theological never entertained its existence and meaning outside theological questions? When intellect became the center of attention, the Renaissance came about. Who would believe that even when the intellect occupied so much of the attention of people and reason even became a deity, it was not perceived that each individual was a social being? Once this awareness reached the universe of relationships between human beings these became utterly different from what they had been. Societies became realities and their study the most important task of all. Who would believe that man enlightened by reason, moved by compassion, involved in changing society, had not seen the person in every individual – had not even seen that there was need for the notion of a person?

History and literature are there to testify to these metamorphoses of the centers of interest of groups of curious people in the Western world, which took place at certain dates, brought about by the sight of specific individuals who became the leaders of their times and for that evolution.

It is as if the vision of reality is blurred by the dedication of the self to some absolute. Reality as a construct, is given more truth by the act of some individual awareness. The dialectic between truth and reality and reality and truth inspires the onlookers who cease to believe in the absolute and allow themselves to be moved by the new. In turn, the next generation elevates into an absolute the new sight. Then all problems become, for example, social problems, for which people seek only social solutions.

Impasses are met in all these evolutions and each time a new departure is required. Sought first in the same layer again and again, these solutions elude one and despair sets in, until someone sees the obvious that had escaped notice because of familiarity, and proposes a transcendental movement towards another awareness as the answer that meets the questions best.

Much of what was there to be learned becomes obsolete suddenly and the evolution requires the clearing from memory of all that seemed important earlier but has now lost its importance. Social men do not entertain the proofs of the existence of God, which was the most important question a few generations ago.

Evolution carries with it torches that light up the fields which were in the dark at the same time as it returns to the dark the areas that were lit before. Evolution affects everything including memory. To live in a certain layer of awareness calls for some storage of functionings so that they can be mobilized at once when the self needs them. If the stresses in the new layer differ from those in the first, memory has to be overhauled in the way an exile may have to acquire a new language when he finds his language inoperative in the new context.

***

Vertical Evolutions Seen as One Evolution - That of Awareness of Energy at Work

Today when awareness of awareness seems the most urgent act for one's self, we can see evolution becoming the instrument man uses to know himself, climbing from one awareness to another until it becomes clear that the energy of the self generates one's own evolution. Today, in contact with the energy of the self commanding all acts including the slightest variation in one's thought, each of us can see that man has reached the point where the most primitive reality of the four realms, energy, is at work everywhere and works best when the minutest amount yields the greatest returns.

Let us look at this stage of our evolution.

Taking all the horizontal evolutions together the two features we worked on at length are visible: horizontal evolution leads to impasses and on top of the set of these evolved objectifications another evolution becomes possible generating another layer ready for another horizontal evolution leading to another impasse. Thus all vertical evolutions can be taken together too and an evolution of evolutions emerges. Its characteristic: the ways energy functions, always coming closer to itself and now reaching the stage where the reality we contemplate is made only of the subtle movements of energy within energy. All the successful attempts of the working energy still obtain and can develop further horizontally but man's awareness has reached the whole spectrum of energy and is now mainly concerned with the minutest shifts and their consequences.

An illustration may help. To our ears music is made up of energy that is distributed in certain ways on the time line giving each piece a unique profile of energy distribution. To our affectivity music is perceived as the way the incoming energy triggers the re-distribution of residual energy in the self to accompany what reaches us and to prolong its impacts. To our self attentive to energy itself, time has gained its reality capable of being perceived beyond the features of the piece as being that which has to be modulated for music – the music – to exist. The energy of time is no longer quantitative, it is supplied simultaneously by the self knowing itself as time beyond the functions of living at all levels, and by the induction of the music beyond all perceptible features. This happens in the way electromagnetic induction operates between two circuits and the passing currents in them. A structured time (a piece of music) affects a time available to be structured in a listener, leaving behind the latter in the individual when the first has ceased to send out energy capable of affecting the energy system that the individual is. At that time, for the individual, the potential for restructuring time in the manner in which it was received exists and can be contemplated. It is in that realm that we now are: virtually potential energy is to be worked on rather than the actual one although all the systems exist that can actualize the virtual. The self's realm is that one. Human living in that realm becomes the constant entertainment of the most subtle reality because of the self's presence. Presence and sensitivity are two of the attributes of the self in that realm. Presence as sensitivity of the state of the energy. Sensitivity as presence of the self in a state of being that can be upset by the slightest imbalance in the universe where one knowingly dwells.

In that realm of energy quality is the characteristic and quantity is a quality in which "more" of the energy is associated. Quality can also be seen as quantity, but so minute that all measuring instruments fail to register it except the self that is energy, knows itself as energy, and makes itself susceptible to notice the slightest movement of it in oneself.

In that realm which is in contact with all the other realms it becomes possible to live two kinds of lives. One in which everyone of the dynamics connected with each of the previous evolutions can be re-examined so as to absorb the excesses of energy deposited in the acts of living horizontally and to leave to the functionings exactly enough energy to function under the commands of the will and no longer by their momentum as before. In such a state the will too uses extremely minute variations in the residual energy to obtain obedience.

The other kind of life is directed towards the future or the present, but no longer to the past. In it what is attempted is no longer conditioned by what is at work in the previous realms and the previous evolutions. Now the self gives itself to the contemplation of what can be done without expenditure of energy at the level where vulnerability seems beyond all thresholds. In that realm the self resides in the human cosmos as the most sensitive receiver that can probe any change in the universe in which it has tuned in. No one has to turn a button to tune in. The self does it simply by opening up to a source of energy, i. e., to anything or anyone. Sensitivity, or that attribute of the self that lets it relate to all emitters, becomes the property of the self that characterizes the new realm. To live humanly then is to live in awareness connected with one's sensitivities and to give one's time to doing with oneself what is required to find oneself constantly more vulnerable, constantly moving towards more subtle universes where "nothings" are perceived and given value of existence.

In such a universe the four realms of previous evolutions are no longer conditionings. They obey. In fact in the fourth realm the evolutions of the other three realms did obey and a human life was possible because the will commanded the psyche, the psyche the soma and the soma matter. Now all that pre-humanity did with consciousness in order to humanize creation can be revised and re-viewed and made to take its place in the continuous evolution of energy towards the meeting of itself as the most pervasive ingredient of all the universes. Behind the appearances – that also are energy – we needed to meet the reality of the dynamics of energy at work and entertain it as the ultimate reality. That the dynamics of energy are energy results from the self's susceptibility of perceiving it.

Beyond the dynamics there is the probe which perceives it, which can be mobilized to be it while it is also not it and in the awareness of the self aware we enter a new realm of awareness in which the accumulation of distinctive attributes interchangeable one into the other, keeps all things together and separate, transformable one into the other, all real, all true, yet neither necessary nor arbitrary. In that universe in flux the self recognizes itself as being nothing else than itself, free energy source of all forms, all realms in oneself.

The realm of free energy will describe itself by sorting out the human challenges bestowed onto it by being energy and free. In a universe where non-contradiction has vanished, where identity is no longer a problem let alone an axiom, the logic of life will for once be permitted to operate and human beings meet each other for what they are.

That realm is one in which evolution just means living, and living at the human level means evolution all the time. The previous realms do not pose problems nor the people who identify with any one of the previous permissible levels of evolution and of awareness.

In that realm all that which seemed impossible until now may turn out to be easy and commonplace.

Memory in that realm is simply the feeling that one has done something, gone through something and this is re-cognized. The question of how it is possible to be at that level of consciousness and still in one's soma, with one's psyche and one's intellect, with people moved by appetites, is not one for a person who knows that being in that state is open to all who do what is required to give their self its place. Once there we see everybody as the energy capable of being aware of itself and capable of taking their evolution into their hands.

Human evolution as it sees itself today is producing a human earth for human beings to live as cosmic beings beyond all previous stations recorded in the three realms and in history.

Memory is no more than the objectified world, Nor less.

www.ingramcontent.com/pod-product-compliance
Lightning Source LLC
Chambersburg PA
CBHW080549170426
43195CB00016B/2725